Molluscan Faunas of the Gulf of Mexico

Recent biodiversity studies, reported here for the first time, have shown that the molluscan fauna of the Gulf of Mexico is far richer and more complex than previously thought. As a result of these new discoveries, the Gulf malacofauna is shown to contain large numbers of endemic species that reside within four separate biogeographical subdivisions of the larger Carolinian Molluscan Province: the Floridian, Suwannean, Texan, and Yucatanean Subprovinces. These four Gulf biotic components, with each supporting its own endemic fauna, are shown here to be separated by distinct ecological and oceanographic barriers. The resultant physical and genetic isolation has led to the evolution of spectacular sibling species radiations, many unknown and undescribed until now. Some of the most conspicuous and important of these are found in the gastropod families Fasciolariidae, Volutidae, Conidae, Muricidae, and Busyconidae, all of which are dominant predators in their respective benthonic ecosystems. The species within these ecologically-important families, along with hundreds of endemic taxa in 50 other gastropod and bivalve families, are illustrated here on 132 color plates and are discussed in detail in the individual chapters. Special attention is given to the mollusks of poorly studied and virtually unknown ecosystems such as those on the deep reefs off the Florida Keys and Dry Tortugas, the deep water coralline algal beds off western Florida, the Flower Garden Reefs off Texas, the petroleum seeps and brine pools of the Sigsbee Escarpment, the Campeche Bank Archipelago, and the deep water areas at the mouth of the Yucatan Channel.

This new book is unlike previous taxonomic surveys of the Gulf of Mexico mollusks in that it highlights only the endemic species and genera and does not cover the large number of widespread Carolinian and Caribbean taxa that occur with them. In this aspect, the book is designed to be an augmentation to previous faunal surveys, adding hundreds more taxa that had been missed in these older surveys or were described after those earlier works had been published. The emphasis on endemic species and species complexes is meant to underscore the special nature of the Gulf of Mexico malacofauna, setting it aside from all others in the Tropical Western Atlantic Region.

Molluscan Faunas of the Gulf of Mexico

Endemism in North America's Inland Sea

Edward J. Petuch
David P. Berschauer

CRC Press
Taylor & Francis Group
Boca Raton London New York

CRC Press is an imprint of the
Taylor & Francis Group, an **informa** business

First edition published 2025
by CRC Press
2385 Executive Center Drive, Suite 320
Boca Raton, FL 33431

and by CRC Press
4 Park Square, Milton Park, Abingdon, Oxon, OX14 4RN

CRC Press is an imprint of Taylor & Francis Group, LLC

Library of Congress Cataloging-in-Publication Data
Names: Petuch, Edward J., author. | Berschauer, David P., author.
Title: Molluscan faunas of the Gulf of Mexico : endemism in North America's inland sea / Edward J. Petuch and David P. Berschauer.
Description: First edition. | Boca Raton, FL : CRC Press, 2025. | Includes bibliographical references and index. | Summary: "Recent biodiversity studies, reported here for the first time, have shown that the molluscan fauna of the Gulf of Mexico is far richer and more complex than previously thought. The Gulf malacofauna is shown to contain large numbers of endemic species. These species are illustrated here on 132 color plates and are discussed in detail in the individual chapters. The emphasis on endemic species and species complexes underscores the special nature of the Gulf of Mexico malacofauna, setting it aside from all others in the Tropical Western Atlantic Region"-- Provided by publisher.
Identifiers: LCCN 2024024984 (print) | LCCN 2024024985 (ebook) | ISBN 9781032883588 (hbk) | ISBN 9781032886879 (pbk) | ISBN 9781003539117 (ebk)
Subjects: LCSH: Mollusks--Mexico, Gulf of.
Classification: LCC QL419 .P48 2025 (print) | LCC QL419 (ebook) | DDC 594--dc23/eng/20240726
LC record available at https://lccn.loc.gov/2024024984
LC ebook record available at https://lccn.loc.gov/2024024985

ISBN: 978-1-032-883588 (hbk)
ISBN: 978-1-032-886879 (pbk)
ISBN: 978-1-003-539117 (ebk)

DOI: 10.1201/9781003539117

Typeset in Times New Roman font
by KnowledgeWorks Global Ltd.

Dedication

This book is dedicated to the following:

Linda J. Petuch, Eric and Rasa Petuch,

Brian Petuch and Kendra Berentsen, and Jennifer Petuch

and

Felicia Weisbrot Berschauer; Morgan, Jeremy, Lincoln, and Brantley Coker;

Jonathon, Tawni, Nora, and Emmett Berschauer

and

to the memory of Rachel L. Carson

Fasciolaria tulipa (Linnaeus, 1758) form *scheepmakeri* Kobelt, 1875, length 155 mm. Within the Gulf of Mexico, this beautiful, heavily-sculptured variant is found primarily off the Florida Keys.

MOLLUSCAN FAUNAS OF THE GULF OF MEXICO
Endemism in North America's Inland Sea

Table of Contents

Figure 0.1 View of a violent storm approaching Camp Lulu Key, Ten Thousand Islands, Collier County, Florida. Scenes like this are typical of the coast along southwestern Florida and the entire Gulf of Mexico during the summer months.

ACKNOWLEDGMENTS

We thank the following individuals for their help in acquiring important specimens for our research and for their help with our field work. Without them, this book could never have been produced: **Capt. Craig Daniels and Brandon Daniels**, Chokoloskee, Florida; **A. Kenneth ("Kenny") Brown**, Chokoloskee, Florida; **Jimmy Kidder**, Chokoloskee, Florida; **Capt. Houston Brown**, Chokoloskee, Florida; **Brian Ward**, Jupiter, Florida; **Damien Schneider**, Athens, Ohio; **Morris A. ("Andy") Foster**, Jupiter, Florida; **Robert L. Eason**, Paris, Tennessee; **Michael Bruggeman**, Snellville, Georgia; **William Bennight**, Wilmington, North Carolina; **Greg Curry, Sr.**, Key West, Florida; **Robert Owens**, Boca Raton, Florida; **Capt. Dale Bittner**, Key West, Florida; **Patrick Wilson**, Lake Worth, Florida; **Robert Pace**, Miami, Florida; **Douglas N. Shelton**, Mobile, Alabama; **Bernie Johnson**, Ensenada, Mexico; **Kevan and Linda Sunderland**, Plantation Key, Florida; **Theodore Kalafut**, Largo, Florida; **Donnie Benton**, Fort Myers, Florida; **Michael Schenk**, Littleton, Colorado; **Stephen Tressel**, Jupiter, Florida; **Luis Vela**, Merida, Mexico; **Kirk Anders**, Ft. Lauderdale, Florida; **Louis Kotora**, St. Petersburg, Florida; **Mark Johnson**, Wilmington, North Carolina; **Thomas and Paula Honker**, Delray Beach, Florida; **Jeffrey Whyman**, Boynton Beach, Florida; **Albert Deynzer**, Sanibel, Florida; **Eddie Matchett**, Okeechobee, Florida; **Clifford Swearingen**, Ocala, Florida; **Randy Allamand**, Asheville, North Carolina; **Andre Poremski**, Washington, DC; **Marcus Coltro and Jose Coltro**, Sao Paulo, Brazil, Bergeggi, Italy, and Miami, Florida; **Kevin Monsecour**, Bekkevoort, Belgium; **Paul Lamb**, Sarasota, Florida; **Dr. Ronald and Mary Jo Bopp**, Sarasota, Florida; **Dr. Myroslaw Harasewych**, Washington, DC; **Robert F. Myers**, Wellington, Florida; **Darwin Alder**, Houston, Texas; **Dr. Anton Oleinik**, Florida Atlantic University, Boca Raton, Florida; **Felix Salas**, San Jose, Costa Rica; **Dr. Jack Freeman**, Boca Raton, Florida; **Gilberto Naal Mora**, Celestun, Mexico; **Kevin Haley**, Playa del Carmen, Mexico; **Dr. Kenneth and Wendy Keaton**, Lauderhill, Florida; **Linda Zylman**, Old Fort, North Carolina; **Nicholas Mazzoli**, Miami, Florida; **Shawn Webster**, Jupiter, Florida; **Dennis Sargent**, Mount Dora, Florida; **Everett Long**, Wilmington, North Carolina; **Lyle Therriault**, Wilmington, North Carolina; **Dr. Emilio Garcia**, Lafayette, Louisiana; **Dr. Douglas Biggs**, Texas A&M University, College Station, Texas; **Frank Frumar**, Cudjoe Key, Florida; **Dr. Manuel Tenorio**, Cadiz, Spain; **Scott Sommerlatte**, Everglades City, Florida (for the exceptional photos of Texas molluscivorous shorebirds). Special thanks go to **Dr. Marvin and Patricia Glickstein**, North Palm Beach, Florida and **Linda Glickstein**, for the donation of large collections of rare deep water species that they collected during their private dredging trips off southeastern Florida; to **Louie Rundo**, Broadview Heights, Ohio, for the donation of several rare Yucatan species; to **Keith Spring** and **Continental Shelf Associates, Inc.**, Stuart, Florida, for their donation of important new Gulf species; and to **Dr. Mary Wicksten and Heather Prestige**, Texas A&M University Biodiversity Research and Teaching Collections (BRTC), College Station, Texas, for the loan of a new cone shell from the Flower Garden Reefs; **Dwight L. Daniels-Brown**, Chokoloskee, Florida, for the kind donation of the *Scaphella junonia* and *Macrocypraea cervus* specimens shown in this book. For technical assistance, manuscript review, and editorial guidance when assembling this book, we thank **Alice Oven**, **Michele Dimont,** and **Shikha Garg**, CRC Press, Taylor & Francis Group.

The widespread Carolinian Province muricid, *Chicoreus dilectus* (A. Adams, 1855), one of the most beautiful shells found in the Gulf of Mexico. Compare this species with its deep water relative *Chicoreus rachelcarsonae* Petuch, 1987 (Figure 4.22).

INTRODUCTION. The Gulf of Mexico: North America's Inland Sea

The large, semi-enclosed body of water that we refer to as the Gulf of Mexico is actually an ancient sea whose origins can be traced back to the early Triassic Period, over 200 million years ago. Prior to that, in the late Permian Period (approximately 250 million years ago), the area that is now the Gulf of Mexico was a bleak, burning desert within the equatorial region on the supercontinent of Pangaea. Later in the Triassic, huge rift zones developed all across Pangaea, resulting in the splitting and separating of the supercontinent into two large subsidiary continents; Gondwana in the south and Laurasia in the north, and several smaller landmasses scattered across the worldwide Panthalassic Ocean. The rift zones, which were originally deep and wide valleys, became flooded with sea water in the late Triassic and early Jurassic Periods, creating narrow oceans and seas between the two newly-formed continents. One of these Pangaean rift zones developed along the southwestern end of the northern continent of Laurasia and later was flooded by sea water to produce a large embayment-like structure; the incipient Gulf of Mexico.

The tranquil marine world of the Gulf that we see today actually formed in hellish conditions of extreme volcanism and tectonic upheaval, as the supercontinent of Pangaea was being torn apart to produce the present-day configuration of continents and seas. By the end of the Jurassic Period (around 150 million years ago), the Atlantic Ocean had already formed, but was composed of long, narrow northern and southern sections, similar to the modern-day Sea of Cortez along Baja California, Mexico. At this time, the accelerated widening of the Atlantic Ocean, driven by rapid sea floor spreading, caused eastern North America to split from northwestern Africa and northern South America, forming a wide central Atlantic area. This oceanic region subsequently fused with the flooded rift zone embayment along southwestern Laurasia to produce the first recognizable Gulf of Mexico. Throughout the late Jurassic and Cretaceous Periods (from 146 million to 66 million years ago), the deep proto-Gulf basin remained connected to numerous shallow epicontinental seas (continent-covering seas) to the north, such as the Cretaceous Inland Sea and the associated Western Interior Seaway, the Mowry Sea, the Pierre Seaway, and the Niobrara Sea. This ancient oceanic seaway complex is now known to have been the evolutionary birthplace for many of the prominent shells found in the modern Gulf of Mexico, particularly the Busyconidae whelks (see Petuch, Myers, and Berschauer, 2015 for details on busyconid evolution).

During the Late Cretaceous Period, the continental shelf and surrounding coastal areas of the Gulf of Mexico dried out due to lowered sea levels. The thriving shallow water marine ecosystems of the Late Cretaceous were then replaced by immense hypersaline lakes and extensive salt flats. These huge salt deposits were later buried by sand and mud when sea levels again rose during the early part of the Cenozoic Era, giving rise to the huge salt and sulfur deposits beneath present-day Louisiana and eastern Texas. Being less dense than the overlying sediments, the buried Cretaceous salt deposits (evaporite minerals) worked their way toward the surface in the form of immense tear-drop shaped bodies that are referred to by geologists as "diapirs". During their passage through the sediment layers, the diapirs actually squeeze petroleum droplets from the oil-bearing geological units, causing pools of oil to accumulate on their upper surface. These oil deposits, sitting on top of rising salt diapirs, are most abundant on the continental shelf off Louisiana and northern Texas and are the targets for the hundreds of oil drilling platforms seen offshore in the northern Gulf. When the diapirs actually intersect the sea floor within the deeper waters of the Gulf, they dissolve and produce water that is so salty that it cannot mix with the normal sea water above it, forming giant hypersaline brine pools that fill depressions on the sea floor. These brine pools are enriched with methane gas and the methane-utilizing bacteria in the water column (methanophilic bacteria) act as a food resource for numerous deep water bivalve mollusks. These bivalves, mostly giant mussels, often form extensive beds that cover acres of the sea floor (these unique deep water Gulf ecosystems are described in Chapter 2 of this book).

Probably the single most significant event to have ever occurred within the Gulf of Mexico region was the impact of a huge asteroid that took place at the very end of the Cretaceous (66 million years ago). This marks the geochronological boundary between the Mesozoic and Cenozoic Eras. At that time, a 10 km-wide asteroid struck the shallow continental shelf area of the southern Gulf embayment, producing a crater over 180 km in diameter and 20 km in depth and centered on what is now the Yucatan Peninsula and the Campeche Bank (Hildebrand, *et al.*, 1991). The present-day town of Chicxulub Puerto, Yucatan State, Mexico sits within the geometric center of the immense crater, on top of a series of buried concentric hills and immense ejecta blankets. The multiple mega-tsunamis that were generated by the impact extended inland as far as present-day Montana and Colorado, flooding and destroying everything within the last remnants of the Cretaceous Inland Sea and permanently altering the Gulf of Mexico shorelines and continental shelves. Even with this extreme catastrophism, many groups of mollusks survived and continued to flourish in the shallow seas of later time periods. These surviving taxa and their modern species radiations will be discussed in the following chapters.

Subsequent to the Late Cretaceous Chicxulub Impact and its accompanying climatic change, the older epicontinental seas became filled with sediments, with only a small oceanographic remnant existing in the form of a narrow embayment in what is now the Mississippi River Valley. During that time, in the Paleogene Period of the Cenozoic Era (Paleocene, Eocene, and Oligocene Epochs, from 66 million to 23 million year ago), this last epicontinental oceanographic remnant developed into the **Mississippi Sea** (sometimes referred to as the "Mississippi Embayment") and contained its own water current systems, deltas, and unique marine ecosystems (discussed in Chapter 2 in this book). These marine communities housed the ancestors of the shells shown throughout this book. As the Mississippi Sea filled with sediments, its unique mollusks, many of them survivors of the Late Cretaceous extinction event, gradually migrated southward into the deeper Gulf of Mexico and evolved into the endemic species radiations that we see today. At the end of the Eocene, around 35 million years ago, a second large asteroid impact produced devastating effects on the Florida Peninsula, at that time only a series of carbonate banks, and the shorelines of the Mississippi Sea. Unlike the Cretaceous impact along Yucatan, the terminal Eocene impact occurred at what is now the mouth of Chesapeake Bay, along present-day Virginia. Immense mega-tsunamis were generated by the initial impact and these travelled down the eastern United States, crossed over the Florida Banks, and spread across the Gulf of Mexico and Mississippi Sea, permanently altering the shape of their coastlines. The huge amount of dust and debris that was blasted into the atmosphere also produced a "nuclear winter" climatic cooling event that led to the extinction of many of the Mississippi Sea endemic mollusks and marine ecosystems. For a detailed overview of the effects of this impact-generated catastrophism on late Eocene Florida, see Petuch and Berschauer, 2021b.

Being enclosed by land on more than three sides, the Gulf truly qualifies as an **Inland Sea**, having only two narrow connections to other bodies of water; the **Yucatan Channel**, connecting the southeastern Gulf to the Caribbean Sea, and the **Straits of Florida**, which connects the southeastern Gulf to the Atlantic Ocean north of Cuba (see Figure 0.2). The Yucatan Channel is shallower than the deep basins on either side, and contains the **Yucatan Sill**, a raised wall-like structure that is around 1000 m below surface and connects the Yucatan Peninsula with western Cuba. Because of its large latitudinal extent, the Gulf of Mexico contains numerous distinct climate-related ecozones, ranging from cool-temperate areas in the north (with air temperatures frequently dropping below 0 degrees C during the winter) to high-tropical areas in the south (with air temperatures averaging around 24 degrees C all year round). This wide range of temperatures, within a single enclosed oceanographic basin has allowed for the evolution of four separate molluscan faunas. In this book, we emphasize how different these four oceanographically-separated malacofaunas really are from each other, illustrating multiple specimens of many of their rarest and least-known taxa.

The new book is also unlike previous taxonomic surveys of the Gulf of Mexico mollusks in that it highlights only the endemic species and genera and does not cover the large number of widespread Carolinian and Caribbean taxa that occur with them. In this aspect, our new treatment is designed to be an augmentation to previous faunal surveys, adding hundreds more taxa that had been missed in these older surveys. The emphasis on endemic species and species complexes is meant to underscore the special nature of the Gulf of Mexico malacofauna, setting it aside from all others in the Tropical Western Atlantic Region. We also illustrate, for the first time in any other book, many newly-described species and newly-rediscovered "lost species" from the Gulf waters, adding to the level of biodiversity that has been documented in the older literature.

With the rapidly-warming worldwide climate, and the resultant increased heating of the Gulf of Mexico waters, many of the endemic Gulf taxa and ecosystems are now becoming endangered and may be lost for all time. In response to these potential extinctions and extirpations, we present our work as a baseline documentation of the spatial distributions of the Gulf mollusks that are present during the early 21st century. The data contained in the following chapters also can be used for comparisons of future patterns of biodiversity with those that exist there now. The sheer magnificence of the endemic Gulf mollusks is also emphasized throughout this book, with large-format photographs of the most important and conspicuous species and subspecies being given in iconographies at the end of each chapter. It is our sincere hope that this endeavor captures the mystery and beauty of the unique molluscan faunas of the Gulf of Mexico, the greatest natural treasures of Eastern North America's inland sea.

Edward J. Petuch, Ph.D. and David P. Berschauer, J.D., 2024

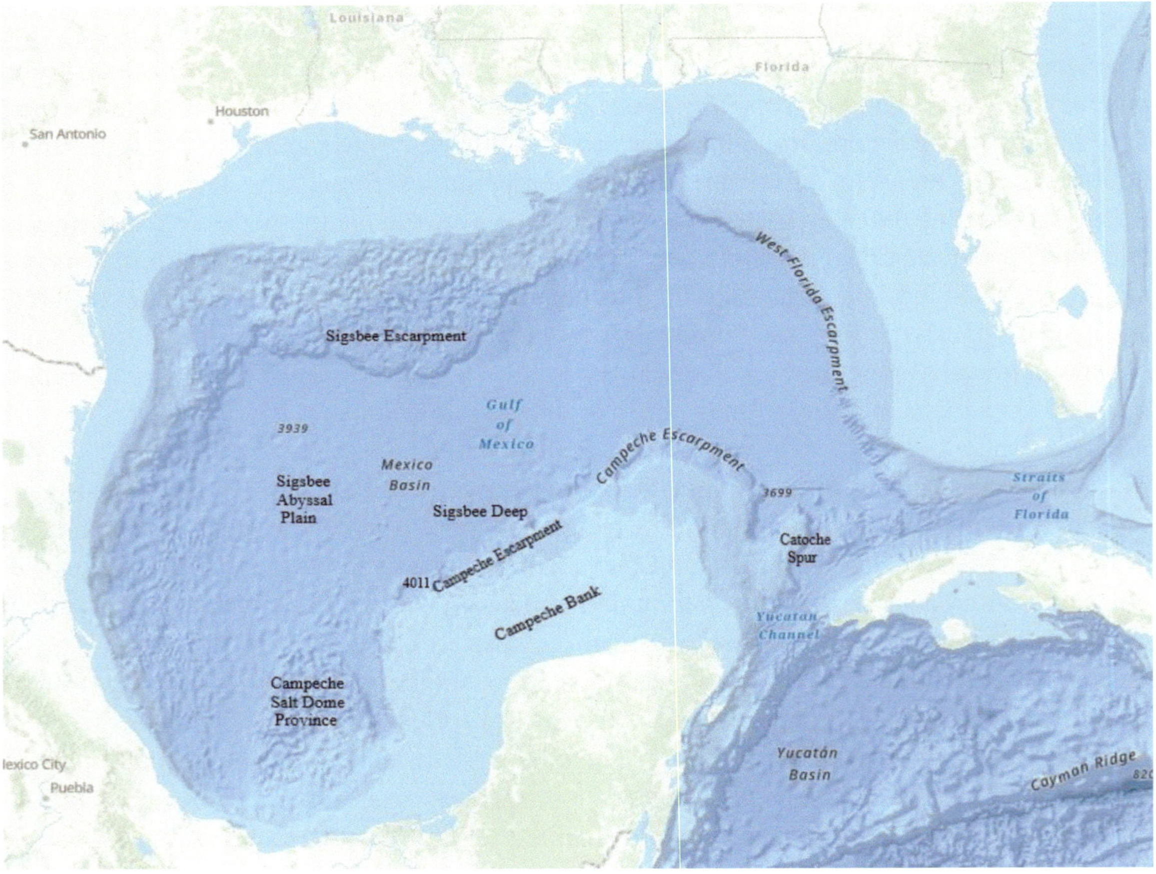

Figure 0.2 Map of the Gulf of Mexico, showing the bathymetric contours and major geomorphological features such as escarpments, terraces, basins, and plains. The abyssal plain of the Sigsbee Deep is one of the most prominent features.

CHAPTER 1. The Carolinian Molluscan Province

Of all the marine invertebrates living in the Gulf of Mexico, the members of the Phylum Mollusca are the most prominent, having the largest biomass and highest biodiversity. Mollusks are so abundant at most Gulf localities that they literally dominate the shoreline and sea floor ecosystems, often overwhelming all other benthonic organisms. All of the shells found in the Gulf inland sea are contained within a single immense marine biogeographical and macroecological unit; the **Carolinian Molluscan Province**. Named for North and South Carolina (see Petuch, 2013; Petuch and Berschauer, 2021a), the Carolinian Province and its fauna ranges from Cape Hatteras, North Carolina, southward along the coasts of South Carolina, Georgia, and eastern Florida, around the Florida Keys, and throughout the entire Gulf of Mexico as far as Cozumel Island, Quintana Roo State, Mexico (Figure 1.1). The Gulf component of the Carolinian Province contains a hybrid fauna, composed of three separate faunal elements: **1.** widespread classic Carolinian species (ranging from Cape Hatteras to Isla Mujeres); **2.** migrants from the Caribbean Province to the south; and **3.** species that evolved locally and are endemic to the Gulf of Mexico basin. These are all discussed in detail in the following chapters of this book.

The presence of numerous classic Caribbean Province index species within the Gulf led several previous workers, such as Warmke and Abbott (1961) and Abbott (1974), to include the Gulf of Mexico within their expanded concept of the Caribbean Molluscan Province (a list of Caribbean species that are present in the Gulf of Mexico is given here in Table 2). Subsequent quantitative analyses of the molluscan faunal data (Petuch, 2013; Petuch and Berschauer, 2021a) have shown that the Gulf malacofauna is distinct from that of the Caribbean, with percentages of endemism surpassing the minimum 50% required for full provincial status and reaching over 70% levels. Besides the high level of endemism seen in the Gulf faunas, all of the classic Carolinian Molluscan Province index species are present within the Gulf, demonstrating that this area represents the southern component of a single widespread eastern North American paratropical province (see Petuch and Berschauer, 2021a for detailed descriptions). Classic Carolinian gastropod index taxa such as the strombid *Strombus alatus*, the cypraeid *Macrocypraea (Lorenzicypraea) cervus*, the muricid *Hexaplex fulvescens*, the fasciolariid *Triplofusus giganteus*, and the conids *Kohniconus delessertii* and *Dauciconus amphiurgus*, and bivalve taxa such as the pectinid *Nodipecten fragosus* and the venerid *Mercenaria campechiensis*, are common throughout the Gulf of Mexico and these large and conspicuous species can be used to define the limits of the Carolinian Province. A list of the classic Carolinian index species is given here in Table 1.

As presently understood, the Gulf of Mexico component of the Carolinian Molluscan Province can be further subdivided into four distinct subprovinces, each with its own suite of endemic species. The four discrete faunas are separated from each other by a series of ecological barriers, generally consisting of intervening areas of lowered salinities (such as the Mississippi River Delta), sediment changes (muddy sediments as opposed to clean carbonate and quartz sands; Figure 1.1), or cold winter currents. Some areas, such as the Texas coast, are characterized by ecological extremes, with low salinities and hot temperatures during the summer and high salinities and cold temperatures during the winter months.

Figure 1.1 View of a classic Carolinian Province coastal environment near Cedar Key, Levy County, Florida. Here, Cordgrass (*Spartina alterniflora*) beds form extensive salt marsh meadows and the Virginia Oyster (*Crassostrea virginica*) grows in small clumps along the low tide line. This type of habitat houses large numbers of characteristic Carolinian taxa such as busycon whelks, *Cinctura* Banded Tulip Shells, and large *Mercenaria* Venus Clams, and is typical of the Suwannean Subprovince intertidal areas along northwestern Florida. This area, part of the Suwannean Subprovince, has cold water temperatures during the winter months, preventing the high tropical mollusks of the Florida Keys from becoming established here.

Gulf Subprovinces of the Carolinian Molluscan Province

Based upon percentages of endemism in their malacofaunas (containing a minimum of 25% endemism at the species level, as defined by Petuch, 2013 and Petuch and Berschauer, 2021a), four distinct subprovinces are now known to exist within the Gulf inland sea (Figure 1.2). These include:

1. Floridian Subprovince: confined to the southeastern Gulf; a high-tropical carbonate and coral reef area encompassing the Dry Tortugas, the Florida Keys, Biscayne Bay, and northward to Palm Beach. The entire Florida Coral Reef Tract is contained within this subtropical subprovince, as are the adjacent deeper water terraces (see Chapter 3).

2. Suwannean Subprovince: confined to the eastern and northeastern Gulf, and extending from the Ten Thousand islands of southwestern Florida, the western coast of Florida and its barrier islands, the emergent grass beds of the muddy northeastern Gulf, the Florida Panhandle sand beaches, to the salt marshes of the Mississippi River Delta. Winter temperatures can drop below freezing, preventing the establishment of tropical marine communities (see Chapter 4).

3. Texan Subprovince: confined to the western Gulf and ranging from the western side of the Mississippi River Delta, down the entire coast of Texas, to near Veracruz, Veracruz State, Mexico. The Texan Subprovince is dominated by long stretches of quartz sand beaches, sand barrier islands, and muddy coastal lagoons. Like the Suwannean area, winter temperatures frequently drop below freezing, preventing tropical marine communities from being established (see Chapter 5).

4. Yucatanean Subprovince: confined to the Mexican coast of the southwestern Gulf and containing high tropical carbonate environments and coral reef areas, the Yucatanean Subprovince is ecologically-separated from the adjacent Texan Subprovince by the fresh water lagoons along the coast of Tabasco State, Mexico and the immense Laguna de Terminos (see Chapter 6).

All four subprovinces are separated from each other by intervening ecological barriers. These inhospitable environments prevent the gene flow between populations, particularly those with direct development, and sequester ecosystems into separate geographical parcels. The strongest ecological barriers present between the subprovinces, and defining their boundaries, are those containing large amounts of fresh water effluent. The largest and most impressive fresh water barrier in the Gulf is the Mississippi River Delta, extending over parts of northeastern Texas, all of Louisiana, and parts of Mississippi State. This physiological barrier separates the organisms of the Texan Subprovince to the west from those of the Suwannean Subprovince to the east. The entire northern half of the Gulf of Mexico also is subject to very cold air temperatures during the winter months, resulting in only temperate water temperatures for half the year. The combined fresh water effluent and low water temperatures prevents many tropical species from migrating into the area. Another important fresh water ecological barrier is seen in the myriad of fresh water lagoons along Tabasco State, Mexico and the huge Laguna de Terminos on the Tabasco-Yucatan State border. The sediment types also shift at this area, going from a muddy quartz sand terrigenous sea floor and beaches (as is typical of the Texan Subprovince) to a carbonate sea floor and extensive coastal mangrove forests.

The smallest and most subtle barrier is present between the southern end of the Suwannean Subprovince and the Floridian Subprovince along extreme southwestern Florida. Here, the fresh water effluent of the Shark River and the Everglades estuaries prevents most of the Florida Keys mollusks from migrating northward into the Ten Thousand Islands area. An abrupt transition occurs along this small stretch of the Everglades coastline, starting just south of Pavilion Key in the Ten Thousand Islands and extending to Ponce de Leon Bay and Cape Sable. Here, the quartz sand beaches and muddy estuarine environments of the Ten Thousand Islands disappear and give way to the carbonate sediment sea floors and coral reefs that typify the Florida Keys and the Floridian Subprovince. Classic Suwannean Subprovince index species such as the olivid *Americoliva sayana sarasotaensis* and the cone shell *Gradiconus floridanus* range only as far south as Pavilion Key and are absent in the Shark River-Cape Sable area. South and east of there, these Suwannean gastropods are replaced by Floridian Subprovince index species such as the olivid *Americoliva bollingi* and the cone shell *Gradiconus burryae*. These taxa will all be discussed and illustrated in the subsequent chapters.

An American Crocodile (*Crocodylus acutus*) swimming in Ponce de Leon Bay, at the mouth of the Shark River, Monroe County, Florida. This river is the biogeographical barrier that separates the Floridian and Suwannean Subprovinces of the Carolinian Province. (Photo courtesy of Morris "Andy" Foster.)

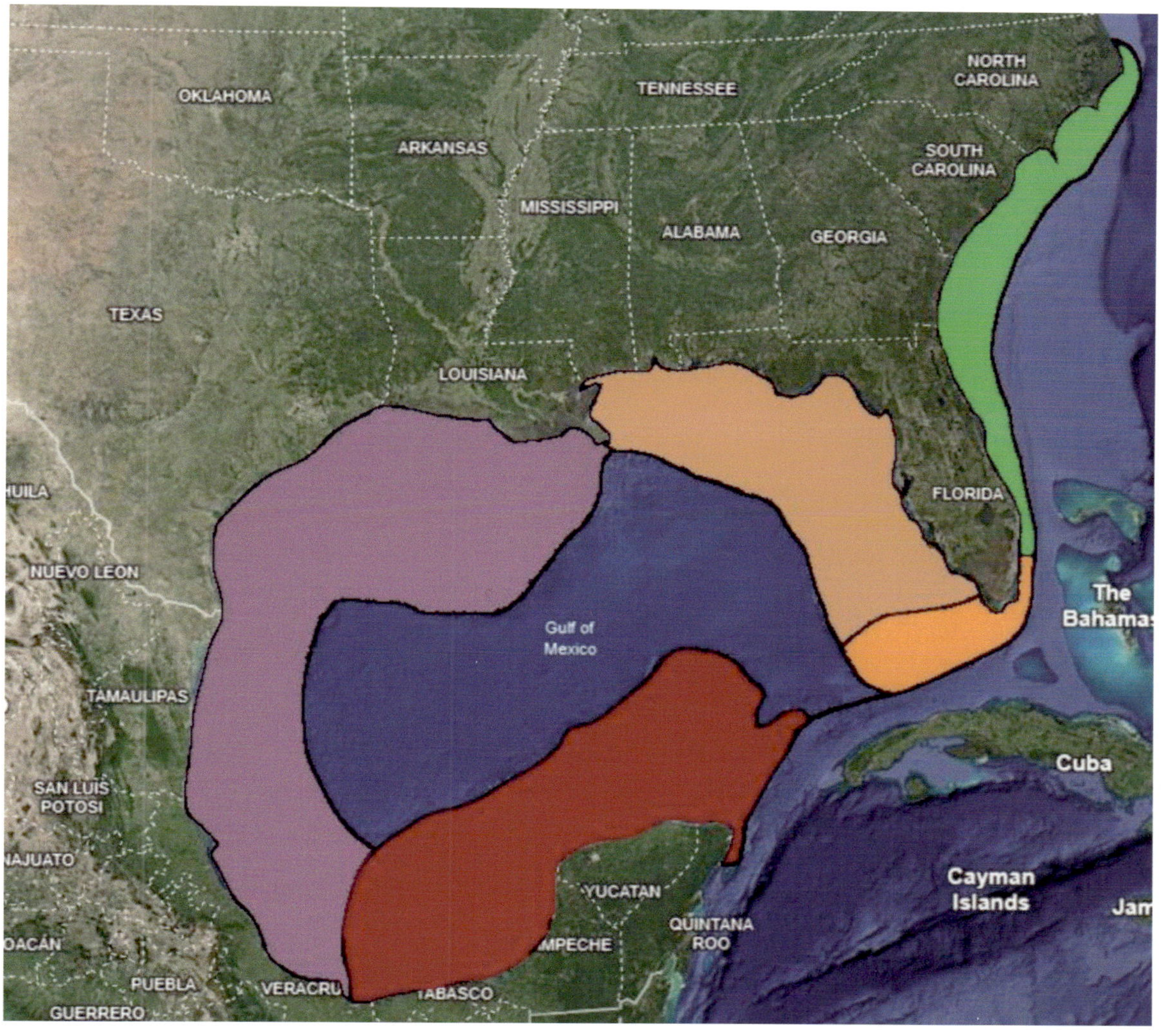

Figure 1.2 General map of the **Carolinian Molluscan Province**, showing its subprovinces. Green= **Georgian Subprovince**; Orange= **Floridian Subprovince**; Brown= **Suwannean Subprovince**; Purple= **Texan Subprovince**; Maroon= **Yucatanean Subprovince**.

The mollusks listed here in Table 1 are considered to be classic Carolinian provincial index species and their combined distributions define the limits of the provincial boundaries. Since most of these well-known taxa are illustrated in many older books, including Tunnell, *et al.* (2010; including micromollusks), Abbott (1974), Petuch (2013), and Petuch and Berschauer (2021a and 2022), we have chosen to illustrate only a few of the more conspicuous and important Carolinian Province index species here (see Figures 1.4-1.11).

TABLE 1. Widespread Carolinian Province Index Species.

Gastropoda

Calliostomatidae

Calliostoma euglyptum (A. Adams, 1855)
Calliostoma psyche (Dall, 1889)

Modulidae

Conomodulus lindae (Petuch, 1987) (Georgian and Floridian Subprovinces)
Modulus kaicherae (Petuch, 1987) (Georgian and Floridian Subprovinces)

Strombidae

Strombus alatus Gmelin, 1791

Cypraeidae

Macrocypraea (Lorenzicypraea) cervus (Linnaeus, 1771)

Ovulidae

Cyphoma mcgintyi Pilsbry, 1939

Cassidae

Cassis spinella Clench, 1944

Ficidae

Ficus papyratia (Say, 1822)

Muricidae

Chicoreus dilectus (A. Adams, 1855)
Favartia goldbergi Petuch and Sargent, 2011 (Georgian and Floridian Subprovinces)
Hexaplex fulvescens (Sowerby I, 1834)
Murexiella levicula (Dall, 1889)
Pteropurpura bequaerti (Clench and Farfante, 1945)
Pygmaepterys richardbinghami (Petuch, 1987) (Georgian, Floridian, and Suwannean Subprovinces)
Timbellus phaneus (Dall, 1889)
Vokesimurex morrisoni Petuch and Sargent, 2011
Vokesimurex rubidus (F.C. Baker, 1897)

Pisaniidae

Hesperisternia multangulus (Philippi, 1848)

Colubrariidae

Bartschia significans Rehder, 1943 (Georgian, Floridian, and Suwannean Subprovinces)

Buccinidae

Liomesus stimpsoni Dall, 1889
Ptychosalpinx globulus (Dall, 1889) (Georgian, Floridian, and Suwannean Subprovinces)

Fasciolariidae

Cinctura hunteria (Perry, 1811) (Georgian, Floridian, and Suwannean Subprovinces)
Triplofusus giganteus (Kiener, 1840)
Triplofusus giganteus form *reevei* Jonas, 1850

Busyconidae

Sinistrofulgur sinistrum (Hollister, 1958) (Georgian, Floridian, and Suwannean Subprovinces)

Volutidae

Clenchina gouldiana (Dall, 1887)
Rehderia georgiana (Clench, 1946)
Scaphella junonia (Lamarck, 1804) (Georgian, Floridian, and Suwannean Subprovinces)

Cancellariidae

Cancellaria reticulata (Linnaeus, 1767)

Conidae

Conasprelloides stimpsoni (Dall, 1902) (= *C. levistimpsoni* Tucker, 2013)
Dalliconus mcgintyi (Pilsbry, 1955)
Dauciconus amphiurgus (Dall, 1889)
Jaspidiconus pfluegeri Petuch, 2004 (Georgian and Floridian Subprovinces)

Kohniconus delessertii (Recluz, 1843)
Lindaconus atlanticus (Clench, 1942) (and banded color form *aureofasciatus* Rehder and Abbott, 1951)
Terebridae
Cinguloterebra floridana (Dall, 1889)
Neoterebra arcas (Abbott, 1954)
Neoterebra onslowensis (Petuch, 1974)
Bivalvia
Pectinidae
Aequipecten glyptus (A.E. Verrill, 1882)
Nodipecten fragosus (Conrad, 1849)
Veneridae
Mercenaria campechiensis (Gmelin, 1791)

Caribbean Faunal Influence in the Gulf of Mexico

Besides containing a majority of Carolinian and subprovincial endemic index species, the Gulf of Mexico fauna also houses a large component of widespread Caribbean species. Within the Gulf region, the Caribbean Province faunal elements fall into three distinct groups: **1.** deep water (lower Neritic and Bathyal Zones) species that are found throughout the Tropical Western Atlantic; **2.** wide ranging shallow water (Littoral Zone) species; and **3.** coral reef-dwelling species or species confined to carbonate environments such as Turtle Grass beds, carbonate mud sea floors, and areas of coral rubble. The most conspicuous and important Caribbean Province species found in the Gulf of Mexico are listed here in Table 2 and these have been documented by Tunnell *et al.* (2010), Petuch and Myers (2014), and Petuch and Berschauer (2023b).

Several deep water taxa that were originally described from the Caribbean region are now known to be found throughout the Gulf, particularly along the outer edge of the wide continental shelves off Florida, Texas, and Yucatan. Some of these ubiquitous deep water taxa include the pleurotomariid *Entemnotrochus adansonianus*, the haliotid *Haliotis pourtalesii*, the cassid *Sconsia grayi*, the epitoniid *Stenorhytis pernobilis*, the muricids *Pazinotus pazi*, *Murexiella hidalgoi*, and *Siratus beauii*, the harpid *Oniscidia dennisoni*, and the cone shell *Conasprelloides villepinii*. These widespread Caribbean species occur together with classic wide-ranging deep water Carolinian endemic species, such as the muricid *Pteropurpura bequaerti*, the colubrariid *Bartschia significans*, the buccinids *Liomesus stimpsoni* and *Ptychosalpinx globulus*, and the cone shell *Dalliconus mcgintyi*, to produce an interesting composite fauna that is unique to the Gulf. In some of the deep areas along the Floridian and Yucatanean reef tracts, other rare widespread gastropods have also been collected, such as the cowrie *Proadusta surinamensis*, the muricid *Babelomurex pacei*, and the ranellid *Ranularia rehderi* (see Petuch and Myers, 2014).

Figure 1.3 Dorsal and ventral views of an 83 mm specimen of *Phyllonotus pomum* (Gmelin, 1791) from Kice Island, Ten Thousand Islands, Florida. This large and impressive muricid gastropod is widespread throughout the Carolinian Province and ranges from North Carolina to Florida and throughout the Gulf of Mexico, including the Flower Garden Reefs of Texas. It is also common in sea grass beds throughout the Caribbean Province, from the Bahamas eastward to the Greater Antilles and southward into the Caribbean Basin as far as Colombia. Compare this widespread *Phyllonotus* species to its rarely-seen endemic Gulf of Mexico relatives; *Phyllonotus whymani* Petuch and Sargent, 2011 (Figure 4.23) and *Phyllonotus mexicanus* (Petit de la Saussaye, 1853) (Figure 6.3).

The shallow water Littoral Zone of the Gulf of Mexico contains molluscan assemblages that are far less homogeneous than those found just offshore in the Neritic Zone. Being exposed to wide tidal fluctuations, extremes in air temperatures, and periods of heavy rainfall and fluctuations in salinity, the intertidal mollusks of the Gulf characteristically have a low biodiversity. Only a few Caribbean species have been able to establish themselves within the Littoral Zone of the Gulf of Mexico and these are all species that can establish themselves on the scattered rocky outcrops and oyster clumps that sporadically appear in all four subprovinces. Some of these include the muricids *Stramonita floridana* and *Stramonita rustica*, the cerithiids *Cerithium atratum* and *Cerithium lutosum*, and the batillariid *Lampanella minima* (common throughout the Gulf but rare along Texas). These hardy Caribbean-western Atlantic gastropods also have to compete with the local intertidal malacofaunas, composed of Gulf endemic species such as *Stramonita canaliculata* and endemic melongenids subspecies, in the surrounding Floridian, Suwannean, and Yucatanean Subprovinces. The combination of detrimental environmental conditions and heavy ecological competition with local endemic taxa explains the dearth of Caribbean intertidal gastropods within the Gulf area. Widespread Caribbean bivalves (many listed in Table 2) fare better on intertidal sand and mud flats and often are major components of the local infaunal communities.

Many of the Caribbean species listed in Table 2 are present only in the Floridian and Yucatanean Subprovinces, where coral reefs, Turtle Grass beds, and mangrove coastlines predominate. These coral reef-dwelling species are also present in a few scattered reef complexes along the Texan Subprovince, such as the Flower Garden Reefs southeast of Galveston, Texas, and the Lobos-Tuxpan Reef System, Isla de Sacrificios, and Isla de Enmedios reef systems off Tuxpan and Veracruz, Veracruz State, Mexico. Some of the most conspicuous Caribbean species that have become established in coral reef and carbonate environments include the strombids *Aliger gallus*, *Aliger gigas*, *Lobatus raninus*, and *Macrostrombus costatus*, the cowrie *Macrocypraea zebra*, the charoniid *Charonia variegata*, the muricids *Siratus consuela* and *Phyllonotus oculatus*, and the cone shells *Chelyconus testudinarius*, *Stephanoconus regius*, *Gladioconus mus*, and *Jaspidiconus mindanus*. Only the Yucatanean Subprovince houses Caribbean migrants such as the melongenid *Melongena melongena*, the harpid *Morum strombiforme*, and the turbinellid *Turbinella angulata* and these are absent from all the other carbonate environments found around the Gulf. For an overview of the widespread Caribbean and western Atlantic Miter Shells (Costellariidae and Mitridae) present in the Gulf, see Garcia (2023).

TABLE 2. Important Caribbean Province Faunal Elements in the Gulf of Mexico.

Gastropoda

Pleurotomariidae

Entemnotrochus adansonianus (Crosse and Fischer, 1861)
Perotrochus amabilis (Bayer, 1963)

Haliotidae

Haliotis pourtalesii Dall, 1881

Turbinidae

Turbo canaliculatus Hermann, 1781
Turbo castanea Gmelin, 1791
Astralium phoebium (Roeding, 1798)

Gazidae

Gaza superba (Dall, 1881)

Cerithiidae

Cerithium atratum (Born, 1778)
Cerithium eburneum Bruguiere, 1792
Cerithium litteratum (Born, 1778)
Cerithium lutosum Menke, 1828

Batillariidae

Lampanella minima (Gmelin, 1791)

Turritellidae

Torcula exoleta (Linnaeus, 1758)

Strombidae

Aliger gallus (Linnaeus, 1758)
Aliger gigas (Linnaeus, 1758)
Lobatus raninus (Gmelin, 1791)
Macrostrombus costatus (Gmelin, 1791)
Strombus pugilis Linnaeus, 1758 (Floridian and Yucatanean Subprovinces)

Xenophoridae

Xenophora conchyliophora (Born, 1780)

Cypraeidae

Luria cinerea (Gmelin, 1791)
Macrocypraea zebra (Linnaeus, 1758)
Naria acicularis (Gmelin, 1791)
Naria turdus (Lamarck, 1810) (a non-indigenous Indian Ocean species, introduced by oil supertankers)
Proadusta surinamensis (Perry, 1811) (Floridian and Yucatanean Subprovinces)

Epitoniidae

Stenorhytis pernobilis (Fischer and Bernardi, 1857)

Ovulidae

Cyphoma aureocinctum (Dall, 1889)
Cyphoma gibbosum (Linnaeus, 1758)
Cyphoma intermedium (Sowerby I, 1828)

Bursidae

Bursa corrugata (Perry, 1811)
Bursa cubaniana (d'Orbigny, 1847)
Bursa (Lampasopsis) thomae (d'Orbigny, 1847)

Cassidae

Cassis flammea (Linnaeus, 1758)
Cassis madagascariensis Lamarck, 1822
Cassis tuberosa (Linnaeus, 1758)
Cypraecassis testiculus (Linnaeus, 1758)
Oocorys bartschi Rehder, 1943

Sconsia grayi A. Adams, 1855
Charoniidae
Charonia variegata (Lamarck, 1816)
Personiidae
Distorsio (Rhysema) clathrata (Lamarck, 1816)
Distorsio (Rhysema) mcgintyi (Emerson and Puffer, 1953)
Distorsio (Rhysema) perdistorta Fulton, 1938
Ranellidae
Gutturnium muricinum (Roeding, 1798)
Linatella caudata (Gmelin, 1791)
Monoplex aquatile (Reeve, 1844)
Monoplex krebsii (Moerch, 1877)
Monoplex martinianum (d'Orbigny, 1847)
Monoplex nicobaricus (Roeding, 1798)
Monoplex parthenopeus (von Salis, 1793)
Ranularia rehderi (A.E. Verrill, 1950)
Septa occidentale (Moerch, 1877)
Tonnidae
Tonna galea (Linnaeus, 1758)
Tonna pennata (Moerch, 1852)
Muricidae
Babelomurex pacei (Petuch, 1987)
Murexiella hidalgoi (Crosse, 1869)
Paziella pazi (Crosse, 1869)
Phyllonotus oculatus (Reeve, 1845)
Phyllonotus pomum (Gmelin, 1791)
Siratus beauii (Fischer and Bernardi, 1857)
Siratus consuela (Verrill, 1950)
Stramonita floridana (Conrad, 1837)
Stramonita rustica (Lamarck, 1822)
Timbellus havanensis (E. Vokes, 1970)
Fasciolariidae
Fasciolaria tulipa (Linnaeus, 1758)
Hemipolygona carinifera (Lamarck, 1816)
Polygona infundibulum (Gmelin, 1791)
Melongenidae
Melongena melongena (Linnaeus, 1758) (Yucatanean Subprovince only)
Melongena melongena form *multispinosa* Dollfuss, 1887 (Yucatanean Subprovince)
Melongena melongena form *salinarum* Bruggeman-Nannenga and Hummelinck, 1986 (Yucatanean Subprovince)
Turbinellidae
Turbinella angulata (Lightfoot, 1786) (Yucatanean Subprovince)
Vasum muricatum (Born, 1778)
Harpidae
Morum oniscus (Linnaeus, 1767) (Floridian and Yucatanean Subprovinces)
Morum purpureum Roeding, 1798
Morum strombiforme (Reeve, 1842) (Yucatanean Subprovince)
Oniscidia dennisoni (Reeve, 1842)
Mitridae
Cancilla antillensis (Dall, 1889)
Nebularia barbadensis (Gmelin, 1791)
Scabricola nodulosa (Gmelin, 1791)

Bellolividae
Jaspidella jaspidea (Gmelin, 1791)
Olividae
Cariboliva scripta (Lamarck, 1811)
Conidae
Atlanticonus granulatus (Linnaeus, 1758)
Attenuiconus attenuatus (Reeve, 1844) (Floridian and Yuatanaean Subprovinces)
Chelyconus testudinarius (Hwass, 1792)
Conasprelloides villepinii (Fischer and Bernardi, 1857)
Gladioconus mus (Hwass, 1792)
Jaspidiconus mindanus (Hwass, 1792)
Kellyconus patae (Abbott, 1971) (Floridian and Yucatanean Subprovinces)
Stephanoconus regius (Gmelin, 1791)
Drilliidae
Cerodrillia thea (Dall, 1883)
Fenimorea moseri (Dall, 1889)
Fenimorea sunderlandi (Petuch, 1987)
Terebridae
Neoterebra dislocata (Say, 1822)
Terebra taurina (Lightfoot, 1786)
Bivalvia
Arcidae
Arca imbricata Bruguiere, 1789
Arca zebra (Swainson, 1833)
Glycymerididae
Glycymeris decussata (Linnaeus, 1758)
Tucetona pectinata (Gmelin, 1791)
Pteriidae
Pinctada imbricata Roeding, 1798
Pteria colymbus (Roeding, 1798)
Pinnidae
Atrina rigida (Lightfoot, 1786)
Atrina seminuda (Lamarck, 1819)
Atrina serrata (Sowerby I, 1825)
Limidae
Ctenoides mitis (Lamarck, 1807)
Ctenoides scabra (Born, 1778)
Pectinidae
Argopecten gibbus (Linnaeus, 1758)
Caribachlamys pellucens (Linnaeus, 1758)
Laevichlamys multisquamata (Dunker, 1864)
Lindapecten exasperatus (Sowerby II, 1842)
Lindapecten muscosus (Wood, 1828)
Spathochlamys benedicti (Verrill and Bush, 1897)
Spondylidae
Spondylus americanus Hermann, 1781
Spondylus ictericus Reeve, 1856
Lucinidae
Anodontia alba Link, 1807
Codakia orbicularis (Linnaeus, 1758)
Phacoides pectinata (Gmelin, 1791)

Chamidae

Chama congregata Conrad, 1833
Chama florida Lamarck, 1819
Chama macerophylla Gmelin, 1791
Chama radians Lamarck, 1819

Cardiidae

Dalliocardia muricata (Linnaeus, 1758)
Laevicardium serratum (Linnaeus, 1758)
Trachycardium egmontianum (Shuttleworth, 1856)

Veneridae

Callista maculata (Linnaeus, 1758)
Petricolaria pholadiformis (Lamarck, 1818)

Macrocypraea (Lorenzicypraea) cervus (Linnaeus, 1771), length 101 mm, from a crab trap set in 10 m depth off Rabbit Key, Ten Thousand Islands, Monroe County, Florida. This is the largest living cowrie in the world and is a classic Carolinian Province index species that ranges from North Carolina to the Yucatan Peninsula of Mexico. An isolated dwarf subspecies, *M. cervus lindseyi* Petuch, 2013, was recently discovered in the Golfo de Guacanayabo in southeastern Cuba.

ICONOGRAPHY OF WIDESPREAD CAROLINIAN INDEX TAXA

Figure 1.4 Widespread species found throughout the entire Carolinian Province.
A= *Modulus kaicherae* Petuch, 1987, width 12 mm, off Palm Beach, Florida; **B=** *Conomodulus lindae* (Petuch, 1987), length 11 mm, off Palm Beach, Florida; **C=** *Heilprinia timessa* (Dall, 1889), length 84 mm, Panama City, Florida; **D=** *Hexaplex fulvescens* (Sowerby I, 1834), length 144 mm, Panama City, Florida; **E=** *Pteropurpura bequaerti* (Clench and Farfante, 1945), length 36 mm, off Tampa, Florida; **F, G=** *Vokesimurex morrisoni* Petuch and Sargent, 2011, length 48 mm, Fort Pierce, Florida (original description reprinted here in the Systematic Appendix); **H, I=** *Vokesimurex rubidus* (F.C. Baker, 1897), length 32 mm, Kice Island, Ten Thousand Islands, Collier County, Florida (compare this species with *V. marcoensis*, Figure 3.5, Chapter 3); **J=** *Calliostoma psyche* Dall, 1889, width 23 mm, 200 m depth off Key West, Florida.

Figure 1.5 *Cinctura hunteria* from the Floridian and Suwannean Subprovinces.
A, B= *Cinctura hunteria* (Perry, 1811), length 62.3 mm, Turtle Key, Ten Thousand Islands, Florida; **C, D=** *Cinctura hunteria* (Perry, 1811), length 63.4 mm, Turtle Key, Ten Thousand Islands, Florida; **E, F=** *Cinctura hunteria* (Perry, 1811), pale orange color form, Kice Island, Collier County, Florida, length 52.3 mm.

Figure 1.6 *Sinistrofulgur sinistrum* from the Floridian and Suwannean Subprovinces, also found along eastern Florida, in the southern Georgian Subprovince.

A, B= *Sinistrofulgur sinistrum* (Hollister, 1958), length 105 mm; **C, D**= *Sinistrofulgur sinistrum* (Hollister, 1811), length 97 mm; **E, F**= *Sinistrofulgur sinistrum* (Hollister, 1811), length 93 mm. All three specimens were collected in the Ten Thousand Islands of southwestern Florida. For illustrations of all the variations of *Sinistrofulgur sinistrum* in the eastern Gulf, see Petuch, Myers, and Berschauer, 2015. Often erroneously referred to as "*Sinistrofulgur contrarium*", which is actually a 3.5 million year-old fossil from the Pliocene formations of Florida and the Carolinas.

Figure 1.7 *Clenchina gouldiana* from the entire Carolinian Province.
A, B= *Clenchina gouldiana* (Dall, 1887), length 54 mm, banded color form, off Key Largo, Florida Keys, Florida; **C, D=** *Clenchina gouldiana* (Dall, 1887), length 61.6 mm, off South Carolina; **E, F=** *Clenchina gouldiana* (Dall, 1887), holotype, length 62.3 mm, off Cape Fear, North Carolina, USNM 83873, showing that *gouldiana* is typically an uncolored shell without distinct bands (*C. bullisi*, a name that has been used by some shell dealers and collectors for typical unbanded *gouldiana*, was never described and is here considered to be a *nomen nudum*).

Figure 1.8 ***Scaphella junonia*** **from the Georgian, Floridian, and Suwannean Subprovinces. A, B=** *Scaphella junonia* (Lamarck, 1804), length 108 mm; **C, D=** *Scaphella junonia* ((Lamarck, 1804), length 110 mm; **E, F=** *Scaphella junonia* (Lamarck, 1804), color form with fused spots, length 121 mm. All three specimens were collected off the Ten Thousand Islands of southwestern Florida, but the species is found throughout the Georgian, Floridian, and Suwannean Subprovinces. The type locality of *S. junonia* is Sanibel Island, Florida.

Figure 1.9 Widespread Cone shells of the Carolinian Molluscan Province.
A= *Dauciconus amphiurgus* (Dall, 1889), length 27 mm, off Panama City, Florida; **B, C**= *Kohniconus delessertii* (Recluz, 1843), length 82 mm, off Cape Canaveral, Florida; **D, E**= *Conasprelloides stimpsoni* (Dall, 1902), length 49 mm, off Key West, Florida; **F**= *Dalliconus mcgintyi* (Pilsbry, 1955), length 68 mm, off Tampa, Florida; **G**= *Jaspidiconus pfluegeri* Petuch, 2004, length 17 mm, Singer Island, Palm Beach, Florida; also found in the Florida Keys.

Figure 1.10 ***Lindaconus atlanticus*** **from all five Carolinian Subprovinces.**
A, B= *Lindaconus atlanticus* (Clench, 1842), length 50 mm, Kice Island, Ten Thousand Islands, Florida; **C, D=** *Lindaconus atlanticus* (Clench, 1842), dark color form, length 48 mm, from Pavilion Key, Ten Thousand Islands, southwestern Florida; **E, F=** *Lindaconus atlanticus* (Clench, 1942), length 72.1 mm, off Cabo Catoche, Yucatan Peninsula, Mexico; possibly a new subspecies; **G=** *Lindaconus atlanticus* banded color form *aureofasciatus* Rehder and Abbott, 1951, holotype, length 66.6 mm, off the Dry Tortugas, western Florida Keys.

Figure 1.11 Widespread Carolinian Province gastropods.
A, B= *Neoterebra onslowensis* (Petuch, 1974), length 32 mm, Missouri Key, Florida Keys; **C, D=** *Neoterebra onslowensis* (Petuch, 1974), length 34 mm, Stock Island, Florida Keys (see Petuch, 1974 for a discussion of the range of this neglected species); **E, F=** *Neoterebra dislocata* (Say, 1822), length 32 mm, Sanibel Island, Florida, for comparison with *Neoterebra onslowensis*; **G, H=** *Cinguloterebra floridana* (Dall, 1889), length 54 mm, 200 m depth off Big Pine Key, Florida Keys (compare this widespread species with the Gulf endemics *C. lindae* and *C. stegeri*); **I, J=** *Cancellaria reticulata* (Linnaeus, 1767), Marco Island, Florida; for comparison with the similar Gulf endemic species *C. adelae* (Chapter 3) and *C. richardpetiti* (Chapter 4).

Figure 1.12 *Fasciolaria tulipa*, a widespread Carolinian and Caribbean Province species.
A, B= *Fasciolaria tulipa* (Linnaeus, 1758), reddish-brown color form, length 124.4 mm, Sanibel Island, Lee County, Florida; **C=** *Fasciolaria tulipa* (Linnaeus, 1758), red color form, length 81.6 mm, Turtle Key, Ten Thousand Islands, Collier County, Florida; **D=** *Fasciolaria tulipa* (Linnaeus, 1758), brown and purple color form, length 69.7 mm, Turtle Key, Ten Thousand Islands, Collier County, Florida. These specimens of the common and widespread *Fasciolaria tulipa* can be used for comparison with the members of the endemic sibling species complex that includes *Fasciolaria bittneri* (Floridian Subprovince; Chapter 3), *Fasciolaria haleyi*, and *Fasciolaria ixchel* (both from the Yucatanean Subprovince; Chapter 6).

CHAPTER 2. Endemic Mollusks of the Gulf of Mexico

As discussed in the Introduction, the molluscan fauna of the Gulf of Mexico contains a large component of endemic taxa, including unique subgenera, genera, and species complexes. Most of these are very different from confamilial taxa found in the adjacent Caribbean Province and the Georgian Subprovince of the southeastern United States. Altogether, these endemics give the Gulf malacofauna its own distinctive appearance, which sets it aside from the rest of the Carolinian Molluscan Province. The faunal composition is so distinctive that it led Valentine (1973: figure 8.7) to consider the Gulf to constitute a faunal province all to itself (the "Gulf Province"). Later quantitative analyses of the faunas of the individual Gulf subprovinces, undertaken by the senior author (Petuch, 2013: 15-20, 47-74), demonstrated that the Gulf, quantitatively, did not reach the level of a full province but was, instead, a composite of four distinct subordinate faunas (referred to here as subprovinces). These geographically-separated subfaunas occur together with a large component of widespread, ubiquitous species that binds the entire Gulf malacofauna into a single large-scale biotic unit. A list of the most prominent widespread bivalves and gastropods that are confined to the Gulf of Mexico is given here in Table 3 and an iconography of many of these is shown at the end of this chapter.

Of special interest within the Gulf of Mexico are the numerous genera and subgenera that have restricted ranges within the semi-enclosed basin and are not found in the adjacent Georgian Subprovince of the Carolinian Province. Some of the more important of these include: the fasciolariid subgenus *Cinctura (Hollisteria)* (see Petuch and Berschauer, 2020), which is restricted to the Yucatanean Subprovince; the busyconids *Lindafulgur*, which is found in both the Suwannean and Yucatanean Subprovinces, and *Busycoarctum*, which is restricted to the Yucatanean Subprovince; the turbinellid genus *Siphovasum*, which is restricted to the Yucatanean Subprovince; and the scaphelline volutids *Aurinia*, which is found throughout the Gulf, and *Garciavoluta*, which is endemic to the eastern Yucatan Peninsula. Of equal interest are several large species radiations of widespread endemic Carolinian Province genera that are restricted to the Gulf of Mexico basin. Some of the more important of these include: the *Melongena (Rexmela)* species complex (Melongenidae), containing three species, two subspecies, and multiple named forms in the Gulf region; the *Cinctura* complex, containing seven species and three subspecies (Fasciolariidae); the *Americoliva* complex, containing nine sibling species (Olividae); the *Scaphella* species complex, containing seven sibling species, and the *Clenchina* complex, with nine sibling species (both Volutidae: Scaphellinae); and the *Gradiconus* species complex, with eight species (Conidae). The members of these sibling species complexes are discussed, individually, within the following chapters of this book.

The Mississippi Sea: Ancestral Home of the Gulf of Mexico Molluscan Fauna

During the late Paleocene Epoch (around 60 million years ago), the entire present-day Mississippi River Valley was flooded with sea water, creating a shallow, triangular sea, with its apex situated in what is now southernmost Illinois. From there, the sea broadened and encompassed the areas that are now western Tennessee, eastern and southeastern Arkansas, eastern Texas, and most of Louisiana and Mississippi. Although referred to as an "embayment" by both geologists and geomorphologists, this elongated, shallow body of salt water actually represented a northern extension of the Gulf of Mexico and is here referred to, in the oceanographic sense, as the **Mississippi Sea** (Figure 2.1). The eastern side of the Mississippi Sea was bounded by the Appalachian Mountains and its western side was bounded by the Ozark and Ouachita Mountains. The sea appears to have formed from the flooding of a failed rift valley (an aulacogen), the result of the North American Plate having passed over a mantle plume "hot spot" during the late Cretaceous Period. This rifting event, and its accompanying grabens, created a depression down the axis of the embayment and later formed a lowered area that now houses the Mississippi River. The failed rifting event, the large-scale graben faults, and the residual heat from the hot spot, all contribute to the seismic activity and thermal springs that characterize areas of the modern central Mississippi River Valley.

During the Eocene (56 to 34 million years ago), the global climate was warmer than it is today and the Mississippi Sea reached its maximum size and development. Extensive rain forests lined the shorelines of the inland sea, palmetto trees grew as far north as southern Illinois, and the first Mangrove Trees and Turtle Grass beds were just beginning to evolve. Large paleo-rivers emptied into the northern half of the Mississippi Sea, depositing huge amounts of sediments that resulted from the erosion of the Appalachian, Ozark, and Ouchita Mountains. These often formed large deltaic complexes and shallow muddy areas, creating the perfect habitats for shallow infaunal predatory gastropods such as scaphelline volutes, busyconid whelks, and large melongenids (examples shown here in Figure 2.1). By the Oligocene, the deltas had become greatly enlarged and extended across all of the areas north of present-day Yazoo City in central Mississippi State. By the beginning of the Miocene Epoch (Aquitanian Age), the entire Mississippi Sea was filled in by sediments and had evolved into expansive grassland areas that supported herds of the first large equine, rhinocerotid, proboscidean, and bovine mammals. The molluscan fauna of the now-infilled Mississippi Sea migrated southward and occupied the shallow banks of the Floridian Peninsula (the Ocala Banks; see Petuch, 2004: 23-34) and the southern coastlines of the Appalachian Peninsula in Alabama, Georgia, and the Florida Panhandle. During the late Oligocene, Miocene, Pliocene, and Pleistocene Epochs, these surviving faunas gave rise to the modern Gulf of Mexico malacofaunas (see Petuch and Berschauer, 2021a; Petuch and Roberts, 2007; and Petuch, 2004 for the paleoceanography, paleogeography, and paleoecology of the Florida area).

One of most prominent gastropod families found in the Gulf today is the Busyconidae; not only in abundance and size, but also as important primary predators within their respective ecosystems. The Family Busyconidae is a totally eastern North American endemic group, having first appeared in the late Eocene Mississippi Sea as the small species *Archefulgur fulguriparens* (see Petuch, Myers, and Berschauer, 2015 for an illustrated overview of busyconid evolution). By the Oligocene, this primitive ancestor had evolved into a complex of closely-related species belonging to the genus *Spinifulgur* (with species like *Spinifulgur spiniger* shown here in Figure 2.3 A, B). By the Miocene and the infilling of most of the Mississippi Sea, the Family Busyconidae had evolved a large number of new genera and subgenera and spread across the submerged Florida Peninsula to the south and up the eastern coast of the United States as far north as New Jersey. The Oligocene and early Miocene genus *Spinifulgur* gave rise to characteristic late Miocene eastern North American genera such as *Sycopsis*, *Turrifulgur*, and *Coronafulgur* (Petuch, Myers, and Berschauer, 2015). These prominent late Miocene genera, in turn, underwent a massive adaptive radiation during the Pliocene and gave rise to advanced genera such as *Busycon*, *Sinistrofulgur*, *Brachysycon*, *Fulguropsis*, *Pyruella*, *Laevisycon*, *Lindafulgur*, *Busycoarctum*, and *Busycotypus*, all of which occurred together and dominated the rich neritic ecosystems of the Floridian Peninsula (Petuch and Berschauer, 2021b). During the numerous Pleistocene climate-driven extinction events, the genera *Brachysycon*, *Pyruella*, and *Laevisycon* disappeared from the fossil record while the other six genera have persisted into the present. Of these, *Busycoarctum* and *Lindafulgur* are now found only in the Gulf of Mexico while *Busycon* is now restricted to the eastern coast of the United States, from Cape Cod to Cape Canaveral.

Another major group of gastropods that is prominent within the Gulf is the Subfamily Scaphellinae, Family Volutidae. Like the busyconids, the scaphelline volutes evolved within the Mississippi Sea during the Eocene and migrated southward as the inland sea filled with sediments during the late Oligocene. Of these endemic volutes, the inflated, bubble-like genus *Caricella* dominated the late Paleocene and Eocene Mississippi Sea (two species shown here in Figure 2.3) and is found in most Eocene beds that are exposed in Mississippi and Alabama. Mineralogical ultraviolet light photographs of *Caricella* species, such as *C. subangulata* and *C. polita* (from the Moody's Branch Formation; see Dockery, 1977, plate 12, figures 2, 3, 5, 6, and 9), demonstrate that shells of this genus exhibited a color pattern composed of large, dark, checkered spots arranged in rows, exactly like living species of *Scaphella*, *Clenchina*, *Aurinia*, and *Rehderia*. The Gulf scaphelline volutes, the direct descendants of the Eocene *Caricella* species complex, obviously inherited the distinctive checkered color pattern from their ancient ancestors. Other ancestors of the living Gulf volutes, such as the Oligocene genera *Atraktus* and *Reticulacella* (Figure 2.3), also had checkered spots and closely resembled small versions of modern *Scaphella* species. The rich fauna of endemic scaphelline volutes seen today in the deeper waters throughout the Gulf can be considered to be a living representative of the exceptionally-rich Mississippi Sea volute fauna.

Another classic Mississippi Sea genus that has survived within the Gulf of Mexico is the turbinellid *Siphovasum* (as *S. latiriforme*; see Figure 6.16 in Chapter 6). This elongated shell, with its extended siphonal canal and extremely elevated and protracted spire, is now found only along the Yucatan Peninsula of Mexico (an index species for the Yucatanean Subprovince). A very similar, extremely elongated turbinellid, "*Vasum*" *humerosum*, was described from the late Eocene Moody's Branch Formation of Louisiana and Mississippi. Having a comparable size and shape, there can be little doubt that the Eocene *humerosa* is the direct ancestor of the living *latiriforme,* and that *Siphovasum* is another example of a relictual species that found a refuge in the Gulf of Mexico after the drying-out of the Mississippi Sea (see Dockery, 1977, plate 11, figures 5A and 5B for illustrations of "*Vasum*" *humerosum*). No other fossil *Siphovasum* species are known from the rich late Miocene, Pliocene, or Pleistocene beds of Florida, indicating that the genus was already a relictual taxon by that time and has remained in the Yucatan area since the Oligocene.

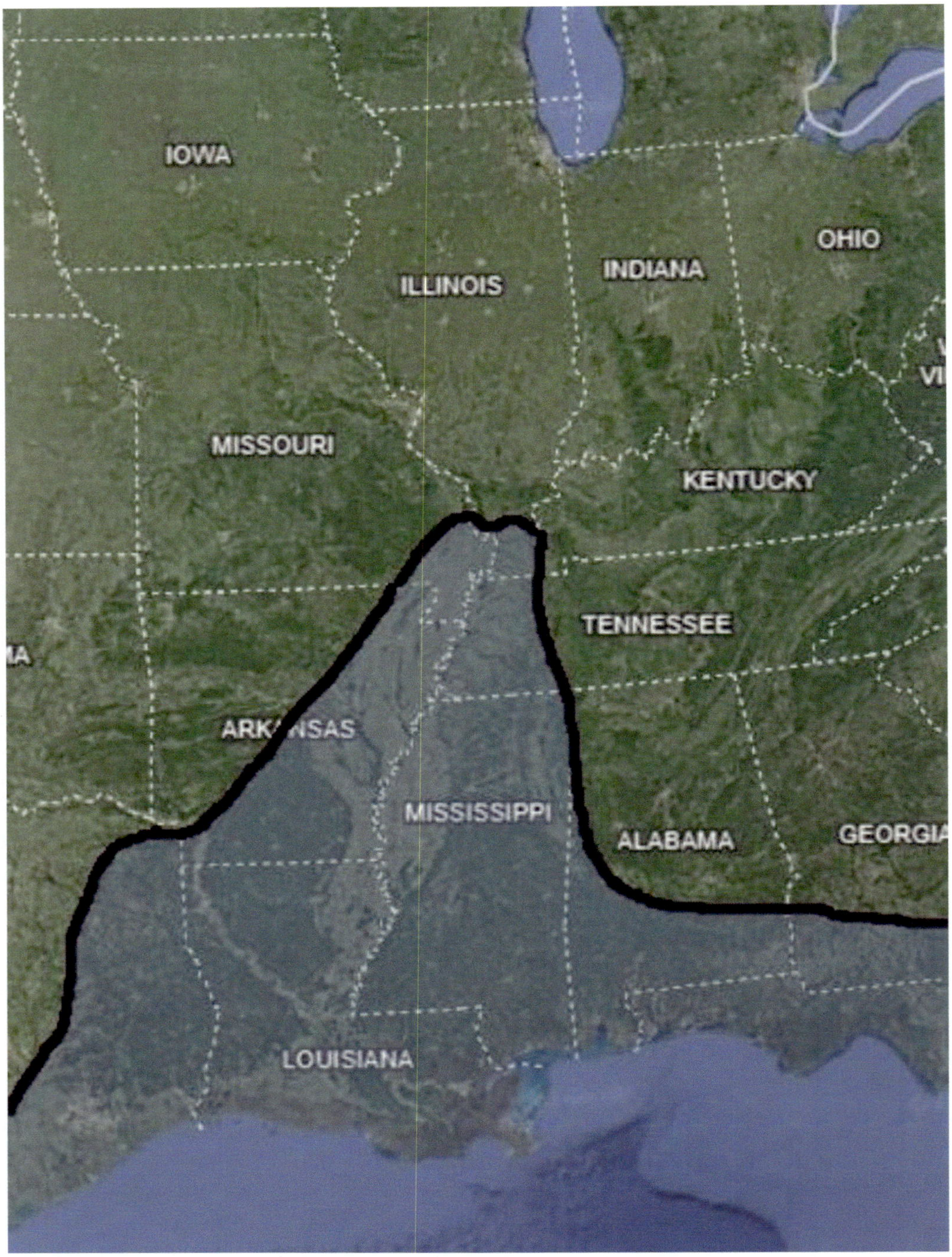

Figure 2.1 Outline of the Mississippi Sea during its maximum development in the Eocene, superimposed upon a map of the present-day eastern United States. Note the distinct elongated shape of the basin, reflecting its origin as a graben; with a series of central blocks having dropped downward relative to the rift valley faults along the eastern and western side. At its maximum development, this epicontinental sea extended as far north as southern Illinois.

Continental Shelf Environments of the Gulf of Mexico

The Continental Shelf areas of the Gulf of Mexico region encompasses all the sea floors from the Intertidal Zone (Littoral Zone) down to 200 m depth. Oceanographically, the continental shelves surrounding the deep basin of the Gulf of Mexico can be divided into two distinct climatic sections; a temperate water northern half and a tropical water southern half. In the southern Gulf, the Yucatanean and Floridian Subprovinces are completely tropical and contain coral reef complexes, Turtle Grass (*Thalassia testudinum*) beds, and Red Mangrove (*Rhizophora mangle*) forests, and are dominated by carbonate sediments. These areas contain the highest biodiversity found in the Gulf and are composed of composite faunas containing both classic Carolinian species and Caribbean migrants (see Chapters 3 and 6 of this book). In the northern half of the Gulf, containing the Texan and Suwannean Subprovinces, the winter weather of the coastal areas frequently drops below freezing temperatures and the offshore waters become too cold for tropical organisms to survive. For this reason, coral reefs and Red Mangrove forests cannot develop within this section of the Gulf and carbonate environments are nonexistent or only poorly-developed. Because of the large river systems that empty into the northern half of the Gulf, the coastal sediment types are dominated by pure quartz sand or sand mixed with organic-rich mud. The distributions of the Gulf of Mexico Neritic molluscan faunas reflect these sediment types and water temperatures.

Considering that its northern shorelines are subjected to months of cold air temperatures, often with winter ice storms and snow flurries, the Gulf should be much cooler than it actually is. This expected cooling, however, is overwhelmed by the effects of the warm **Gulf Loop Current**, a large, prominent hydrographic feature that enters the Gulf of Mexico through the Yucatan Channel and passes over the **Yucatan Sill** (Figure 1.2, Chapter 1). From there this jet-like warm current flows straight northward toward the area extending from the Mississippi River Delta to the Florida Panhandle. At that point, the Loop Current deflects eastward, and abruptly flows southward down the outer edge of the Florida continental shelf. When the current reaches the area of the Dry Tortugas, it curves to the east and flows into the Florida Straits south of the Florida Keys, where it becomes the **Florida Current**, and ultimately, the **Gulf Stream**. In winter, the southward flow of the Loop Current along western Florida carries colder surface water landward, making the shallow coastal areas of western Florida much colder than areas in the adjacent Florida Keys. Large eddies often develop and break off from the main Loop Current, carrying gyres of warm water westward to areas off the Louisiana and Texas coasts. Some of these eddies are long-lived and bring tropical water temperatures to the Flower Garden Reefs off Texas and the reef complexes off Veracruz State, Mexico. These small gyres also carry long-lived planktotrophic larvae to these shallow submarine banks, seeding them with Caribbean mollusks, corals, and echinoderms that normally would not exist that far north in the Gulf.

Because of the large amount of river effluent in the northern half of the Gulf of Mexico, most of the shorelines of the Suwannean and Texan Subprovinces are composed of sand barrier islands that shelter immense brackish water lagoon systems along their landward sides. In some areas, such as along the Mississippi River Delta and its associated islands and in the quiet, sheltered areas along the northeastern Gulf area ("Big Bend" to Cedar Key) of northern Florida, immense wet meadows of Cordgrass (*Spartina alterniflora*), interspersed by open tidal channels, dominate the muddy shoreline and create habitats for many species of endemic Gulf mollusks (Figure 1.1, Chapter 1). Of special interest in these Cordgrass meadows is the Say's Periwinkle, *Littoraria irrorata sayi*, a Gulf of Mexico endemic subspecies of the eastern United States Salt Marsh Periwinkle, *Littoraria irrorata*. This Gulf subspecies (Figure 2.2), originally named from western Florida, differs from the east coast nominate subspecies in having much finer and more numerous cords on the body whorl and in having a bright orange-colored aperture as opposed to the pale orange and brown aperture seen on typical *L. irrorata*. Salt marsh periwinkles, of any kind, are absent from an area extending from Fort Pierce, Florida to the Dry Tortugas, and including the entire Florida Keys. The species, then, can be seen to have been separated into two populations since the Pleistocene; one ranging from Cape Cod southward to Fort Pierce and the other ranging from the Ten Thousand Islands of southwestern Florida (Petuch and Berschauer, 2022) northward to the Mississippi River Delta and along the entire coast of Texas. The genetic isolation of the two populations has allowed the evolution of the differences in shell sculpture and apertural color that readily separate the two subspecies.

Within the Cordgrass meadows, from Mobile Bay, Alabama south to Cedar Key, Florida (the Sueannean Subprovince), the open narrow tidal channels between the grass beds support large aggregations of the oysters *Crassostrea virginica* and *Ostreola equestris*. These, in turn, offer a hard substrate for several large predatory gastropods, including two subspecies and several named forms of the Crown Conch *Melongena (Rexmela) corona* (discussed later in this book in Chapter 4), the Left-Handed Whelk *Sinistrofulgur sinistrum*, and the large muricid *Stramonita canaliculata* and its ornate variety *haysae*. This large *Stramonita* (Figure 2.5) ranges throughout the entire Gulf of Mexico and is the major drilling predator on the oyster beds within the intertidal area (Littoral Zone). In Texas (Texan Subprovince), the salt marsh tidal channels contain a more impoverished molluscan fauna, completely lacking melongenid species and with the Pulley's Whelk, *Sinistrofulgur pulleyi* replacing the eastern Gulf *S. sinistrum*. For detailed descriptions of the Cordgrass and tidal channel molluscan faunas found along the Texas coast, see Tunnell *et al.*, 2010: 36-38.

Protected lagoonal environments along the northern half of the Gulf often house immense beds of sea grass, usually *Halodule wrightii* (Shoal Grass) in the intertidal areas and *Thalassia testudinum* (Turtle Grass) in shallow subtidal areas. Adjacent to the Cordgrass meadows, these Littoral Zone sea grass beds support the most species-rich molluscan faunas found anywhere along the northern Gulf coastlines. Of particular interest within these submerged meadows are the large shoals of Bay Scallops in the genus *Argopecten*, which are ecologically-important within the sea grass ecosystem but also form the basis for a large commercial scalloping industry. Within the Gulf of Mexico, two geographical subspecies of the Bay Scallop occur; with Taylor's Bay Scallop, *Argopecten irradians taylorae* Petuch, 1987 (Figure 4.15, Chapter 4), ranging from the Florida Keys northward to Mobile Bay and with the Wide-Ribbed Bay Scallop, *Argopecten irradians amplicostatus* (Dall, 1898) (Figure 5.4, Chapter 5), ranging from the western side of the Mississippi River Delta south to the Bay of Campeche, Mexico. A similar biogeographical

Figure 2.2 Close-up view of Say's Periwinkle, *Littoraria irrorata sayi*, crawling on Cordgrass, *Spartina alterniflora*, in a salt marsh near Cedar Key, Levy County, Florida. This grass-dwelling periwinkle is a geographically-isolated subspecies of the United States east coast *Littoraria irrorata* and is found only in the Gulf of Mexico. In the Gulf, it ranges from the Ten Thousand Islands to Texas and is not present in the Florida Keys and southeastern Florida (absent from the Floridian Subprovince). This Gulf endemic subspecies differs from the United States east coast *L. irrorata* in having finer and more numerous cords on the body whorl and in having an orange aperture instead of pale orange-tan and brown. The small potamidid gastropod *Cerithidea turrita*, which is endemic to northwestern Florida, lives in the organic-rich mud at the base of the grasses.

pattern is seen in another sea grass bed mollusk, the Banded Tulip Shells of the genus *Cinctura*; with *Cinctura hunteria* (Perry, 1811) (Figure 1.5, Chapter 1) and its subspecies being confined to the eastern side of the Gulf and with *Cinctura lilium* (Figure 5.5, Chapter 5) being confined to the western half of the Gulf. As seen in the Bay Scallops, the Mississippi River Delta acts as an ecological and genetic barrier to the sets of sibling taxa. The Gulf of Mexico scallops are closely-related to two other eastern United States Bay Scallops; the Concentric Bay Scallop, *Argopecten irradians concentricus*, which ranges from Cape Hatteras, North Carolina south to Fort Pierce, Florida and the Northern Bay Scallop, *Argopecten irradians irradians*, which ranges from Cape Cod, Massachusetts south to Cape Hatteras. The Palm Beach and Broward County coastlines of southeastern Florida have no resident Bay Scallops, and the area appears to be an ecological exclusion zone between *concentricus* and *taylorae* (Petuch, 2013: 17; Petuch and Berschauer, 2021a; Petuch and Berschauer, 2022: 20). The four subspecies of Bay Scallops, including the two east coast taxa and the two Gulf endemic taxa, all have both demonstrable morphological differences (rib count and rib structure) and genetic differences (consistent DNA characters; see Marelli *et al.*, 1997). Further studies may demonstrate that the four subspecies actually represent a complex of closely-related sibling species that evolved during the late Pleistocene.

Another gastropod group of special importance within the Gulf of Mexico is the family Modulidae, which has undergone an adaptive radiation that produced at least eight endemic species and subspecies. These endemic modulid taxa are restricted to discrete areas around the Gulf, occurring in both shallow water intertidal and deeper neritic environments. The Floridian Subprovince houses three species in three genera; the Turtle Grass-dwelling *Trochomodulus calusa* and the deep water *Conomodulus lindae* and *Modulus kaicherae* (see Chapters 1 and 3), which are both associated with outer neritic red algal beds. The Suwannean Subprovince also houses two endemic modulid taxa; the Turtle Grass-dwelling *Modulus floridanus* and the deep water *Conomodulus lambi*, which lives on offshore coralline algal beds (see Chapter 4). The most wide-ranging modulid species in the Gulf is *Modulus danielsi*, which lives in shallow sea grass beds along the Florida Panhandle (western edge of the Suwannean Subprovince) and ranges down the entire coast of Texas and eastern Mexico as far as Veracruz (see Chapter 5) (Tunnell *et al.*, 2010: 133). The Family Modulidae in the Yucatanean Subprovince is represented by two endemic species, *Modulus hunahpu* and *Trochomodulus velai*, both of which live in nearshore sea grass beds (see Chapter 6; at least one more unnamed species also occurs in deep water off the Yucatan Peninsula). Considering that the entire family only contains around 20 species worldwide, it is noteworthy that almost half of the known species are restricted to the Gulf of Mexico. The Florida fossil record also contains many extinct species in the genera *Modulus*, *Trochomodulus*, and *Conomodulus* (Petuch, 1994; 2004), demonstrating that the Gulf of Mexico area has been a "hot spot" for modulid evolution during the past 5 million years (also see Petuch and Berschauer, 2023b).

The most frequently-encountered shoreline environment in the Gulf of Mexico is the high-energy quartz sand beach. These beaches extend seaward beyond the surf zone and incorporate the sand sea floors in shallow water immediately offshore (the **Forebeach area**). Most often exposed to strong wave action, these open beaches and their adjacent forebeach areas can support only a small number of resident mollusks, most of which are infaunal (buried) bivalves such as donacids, tellinids, mactrids, and venerids. The main source of the quartz sand found throughout the Gulf is the myriad of rivers that empty into the Gulf basin, having been derived from the erosion of granitoid rocks in the surrounding mountain ranges, from the Ouchita and Ozark Mountains in the west and the Appalachians in the east. A huge amount of sand also derives from the north, being carried into the Gulf by the effluent of the Mississippi River; having been produced by the erosion of silicate minerals on the Canadian Shield and carried into the northern end of the Mississippi River Valley by Pleistocene glaciers. After entering the Gulf in the Mississippi River Delta region, this quartz sand is carried eastward and then southward by the wave-produced longshore current (see Petuch and Berschauer, 2022 for more details on sediment transport). This transported sediment produces the beautiful white sand beaches of the Florida Panhandle and western Florida and is carried as far south as the Ten Thousand Islands. There, at Cape Sable, the sand fans out into Florida Bay and disappears before it reaches the Florida Keys. The coastlines of Texas and eastern Mexico are also typified by quartz sand beaches and shorelines of this type dominate most of the peripheral areas of the Gulf.

Ranging from the shallow subtidal areas of the Littoral Zone, all the way out to the physical edge of the continental plate at 200 m, the continental shelf area (**Neritic Zone**) houses the widest variety of environments and ecosystems found within the Gulf of Mexico. Some of the continental shelves around the periphery of the Gulf are very wide, forming broad, flat platforms. The best developed of these are the **Florida Platform** off western Florida, the **Texas-Louisiana Shelf** off northern Texas and the Mississippi River Delta area, and the **Campeche Bank** off the northern and western sides of the Yucatan Peninsula of Mexico. Most of the eastern coast of Mexico, from Texas south to the Bay of Campeche, has a very narrow continental shelf that drops off abruptly into the Sigsbee Abyssal Plain. A similar, but smaller, version of this narrow shelf pattern is found off the eastern side of the Yucatan Peninsula, where the shelf drops off abruptly at the mouth of the Yucatan Channel. Although a high level of endemism has evolved on each separate platform area, there are numerous widespread neritic mollusks that do occur on all continental shelf areas (many listed here in Table 3). Some of these ubiquitous offshore Gulf of Mexico mollusks include gastropods such as the muricid *Paziella nuttingi* (Figure 2.9), the large fasciolariid *Fasciolaria bullisi* (Figure 2.8), the cancellariid *Ventrilia tenerum* (Figure 2.8), the cone shells *Conasprelloides cancellatus*, *Dalliconus armiger*, and *Dalliconus rainesae* (all three species shown in Figure 2.16), and bivalves such as the pectinid *Euvola marshallae* (Figure 2.18). This thin, fragile scallop forms immense shoals on the Texas and Florida outer continental shelves and these serve as a food resource for large predatory volutid gastropods such as the widespread *Aurinia dubia* (Figure 2.10) and the northern Gulf *Aurinia kieneri* (Figure 2.11).

The Gulf of Mexico Volute Fauna

The scaphelline volute complex of the Gulf of Mexico is the single most spectacular endemic evolutionary radiation found within the American inland sea. These descendants of the highly-endemic Paleogene Mississippi Sea volute fauna all have a checkered color pattern, a shell character that was inherited from their spotted *Caricella* and *Atraktus* ancestors (some shown here in Figure 2.3). Because of this similarity in color patterns, all of these volutes have been incorrectly referred to as "*Scaphella*" species regardless of the shell shape and size and the form and structure of the protoconch and early whorls (*i.e.* Weaver and Dupont, 1970). Ignoring the shared ancestral color pattern (a plesiomorphic shell character), the differences in the protoconchs and columellar plicae readily separate the "scaphellas" into six distinct groups, which are treated here as full genera. In total, 23 species of volutes are now known from the Gulf of Mexico and these are illustrated in the following chapters of this book. These include:

1. *Scaphella* (sensu stricto, Swainson, 1832). Type Species: *Scaphella junonia*. Members of this genus have proportionally-large, rounded and dome-shaped protoconchs, four strong ridge-like columellar plicae, and early whorls that are heavily-sculptured with strong longitudinal ribs. The checkered spots are large and prominent, even on the early whorls of the spire. Species in this genus include the genotype and six geographically and bathymetrically-separated sibling species (previously considered to be subspecies of *S. junonia* but here considered separate full species).

- ***Scaphella butleri*** Clench, 1953; Campeche Bank, Yucatan Peninsula, Mexico (Figure 6.17).
- ***Scaphella curryi*** Petuch and Berschauer, 2019; Texas coast, eastern Mexico as far south as Veracruz (Figures 5.12 and 5.13). This is the only *Scaphella* species that has distinctly reddish spots instead of the normal dark brown or blackish-brown spots.
- ***Scaphella elizabethae*** Petuch and Sargent, 2011; the Dry Tortugas and the western edge of the Florida Platform (Figure 3.14). The original description is reprinted in the Systematic Appendix at the end of this book.
- ***Scaphella glicksteinorum*** Petuch and Berschauer, 2023; a dwarf taxon found on shallow coral reefs from Palm Beach to Biscayne Bay and the entire shallow reef tract of the Florida Keys; may have either reddish-brown or pale brown spots (Figure 3.15).
- ***Scaphella johnstoneae*** Clench, 1953; Florida Panhandle, Alabama, and Mississippi coasts; most often with a deep yellow-orange or yellow base color (Figure 4.34).
- ***Scaphella junonia*** (Lamarck, 1804); North Carolina to the Florida Keys and western Florida as far north as Cedar Key (Figure 1.8). Eastern U.S. specimens may be a separate subspecies.
- ***Scaphella stimpsonorum*** Cossignani and Allary, 2019; eastern coast of the Yucatan Peninsula, Mexico, in the Contoyan Infraprovince (Figure 6.29).

These seven taxa, altogether, represent a close-knit complex of sibling species that evolved during the Calabrian Pleistocene (2 million years BP). Based on similar shell morphology, the living Gulf complex derives from volutes such as *Scaphella capelettii* Petuch, 1994 and the dwarf *S. seminole* Petuch, 1994 from the Bermont Formation of southern Florida (see Petuch and Berschauer, 2021b).

2. *Clenchina* Pilsbry and Olsson, 1953. Type Species: "*Scaphella*" *dohrni*. Members of this genus have smaller, more elongated, and distinctly cylindrical shells, with three to four columellar plicae. Protoconchs are proportionally large, rounded and distinctly dome-like. On most *Clenchina* species the early whorls are heavily ornamented with a single row of large, elongated rib-like knobs. This is the largest group of scaphelline volutes in the Gulf of Mexico, with nine known species. The Gulf *Clenchina* species include:

- ***Clenchina atlantis*** Clench, 1946; deep water Straits of Florida (Figure 3.29).
- ***Clenchina carrieae*** Petuch and Berschauer, 2023; Straits of Florida (Figure 3.34).
- ***Clenchina dohrni*** (Sowerby III, 1903); Straits of Florida, Florida Keys, Dry Tortugas, and the entire western Florida Platform (Figure 2.12).
- ***Clenchina florida*** (Clench and Aguayo, 1940); Florida Keys (Figure 3.25). This ornate species, with its heavily-coronated early whorls, occurs in the Florida Keys, where it lives together with *C. dohrni*. The northern Cuban and western Bahamian *Clenchina bermudezi* (Clench and Aguayo, 1940) may be a subspecies of *C. florida*.
- ***Clenchina gouldiana*** (Dall, 1887); North Carolina to the Florida Keys, entire Gulf of Mexico to the Yucatan Channel (Figure 1.7). This is the most widespread volute found in the Carolinian Province and it occurs in both banded and unbanded color forms. "*Scaphella" garciai* Bail, 2007, a species that lacks shoulder and spire knobs, has been confused with *gouldiana*. The unique holotype of *garciai* was collected off St. Augustine, Florida and reports of its presence in the Gulf of Mexico are erroneous. As a result, it is not included in this book. Further analyses have shown that *garciai* is not a *Clenchina* at all but belongs in the genus *Bathyaurinia*, a group absent from the Gulf. "*Scaphella bullisi*", a name that has been used for typical unbanded *gouldiana* from eastern Florida, was never formally described and is here considered to be a *nomen nudum* and has no nomenclatural status.
- ***Clenchina marionae*** Pilsbry and Olsson, 1953; western edge of the Florida Platform and Florida Panhandle (Figure 4.19).
- ***Clenchina robusta*** Dall, 1889; Straits of Florida, Florida Keys, and the entire Gulf of Mexico (Figure 2.13).
- ***Clenchina rundoi*** Petuch and Berschauer, 2023; Dry Tortugas, Yucatan Channel, northern Cuba (Figure 3.26).
- ***Clenchina worki*** Coltro, 1998; deep water off northern Texas, Louisiana, Alabama, and Sigsbee Escarpment brine pools and oil seeps (Figure 2.14).

3. *Caricellopsis* Petuch and Sargent, 2011. Type Species: "*Scaphella*" *matchetti*. Shell large and very inflated, bulliform. Protoconch proportionally large, rounded, dome-like. First two postnuclear whorls with two rows of large checkered spots and ornamented with a network of criss-crossed fine raised cords, producing a fenestrate sculpture pattern. Later whorls marked with four wide bands of large, amorphous, zig-zag flammules, usually tan-orange or dark brown in color. Columella with only two poorly-developed plicae. The type species, *Caricellopsis matchetti* (Petuch and Sargent, 2011), was named from a specimen that was supposedly collected in the eastern Gulf of Mexico (Petuch and Sargent, 2011: 150). That species has subsequently been found to have been collected by shrimpers off northern Colombia, showing that *matchetti* is actually a southern Caribbean species and not from the Gulf of Mexico. See the Systematic Appendix at the end of this book for the reprinting of the original description. *Caricellopsis contoyensis* is the only member of the genus that occurs within the Gulf of Mexico.

- ***Caricellopsis contoyensis*** (Emerson, 1979); eastern Yucatan, Mexico and the Yucatan Channel (Figure 6.27).

4. *Aurinia* H. Adams and A. Adams, 1853. Type Species: *Aurinia dubia*. This genus and subgenus contains the largest volutes in the Gulf of Mexico, with specimens of *A. kieneri* often reaching 225 mm in length. *Aurinia* shells are very elongated, thin, and fragile with reduced color patterns composed of only a few scattered brown checkers. The protoconchs are composed of two slender whorls and are tightly coiled, resembling an extruded confectionary dessert. No plicae are present on the columella. Only three taxa are referable to this genus and these include:

- ***Aurinia dubia*** (Broderip, 1827); western Florida Platform, Florida Panhandle, Alabama, and Mississippi (Figure 2.10).
- ***Aurinia dubia ethelae*** Pilsbry and Olsson, 1953; Mississippi River Delta, northeastern Texas (Figure 5.11). This bathymetric subspecies is characterized by having large, prominent, fold-like pleats or ribs on the body whorl. This subspecies is also larger than the nominate subspecies and lives in deeper water.
- ***Aurinia kieneri*** (Clench, 1946); Alabama, Mississippi River Delta, deep water northern Gulf from Texas to western Florida Panhandle (Figure 2.11).

5. *Rehderia* Clench, 1946. Type Species: "*Aurinia*" *schmitti*. Resembling miniature versions of the larger *Aurinia* species, members of *Rehderia* differ in being extremely slender and elongated, with proportionally-larger and more rounded protoconchs. Two very faint and poorly-developed plicae may be present on the columellar area, although they are most often absent, as in the larger *Aurinia* species. Only two species are known and these include:

- ***Rehderia georgiana*** (Clench, 1946); North Carolina to southern Florida, Florida Keys, Dry Tortugas (Figure 3.28).
- ***Rehderia schmitti*** (Bartsch, 1931); Florida Keys, Dry Tortugas, and Tortugas Shrimping Grounds (Figure 3.29).

6. *Garciavoluta* Petuch and Berschauer, 2023. Type Species: *"Scaphella" mcginnorum*. This monotypic genus was originally (Petuch and Berschauer, 2023) considered to be a subgenus of the genus *Aurinia* but is now considered to be a full genus all to itself. Along with the genera *Caricellopsis* and *Aurinia*, the genus *Garciavoluta* contains the largest volutes known from the Gulf of Mexico; resembling *Aurinia* species in shape and shell thinness but differing in having a much larger protoconch that is globose and knob-like and dark brown in color. The adult shell is unpatterned, lacking any checkered spots. One or two rows of tiny brown spots may be present on the first three postnuclear whorls and the siphonal canal is dark orange-brown. The columellar area has two proportionally-large, thin plicae.

- ***Garciavoluta mcginnorum*** (Garcia and Emerson, 1987); restricted to the eastern side of the Yucatan Peninsula of Mexico, western Yucatan Channel (Figure 6.30).

Of all these volutes, only *Scaphella junonia*, *S. glicksteinorum*, and *S. elizabethae* are known to live in shallow water (3-30 m depths) and are the only species that can be regularly collected by SCUBA divers. All the rest live in deeper water along the edge of the Neritic Zone, or in the Bathyal Zone escarpment areas, and can only be collected by dredging from research vessels or deep water shrimpers (searching for "Royal Red" *Glyphocrangon* shrimp). Only five species of scaphelline volutes, *Scaphella junonia*, *Clenchina gouldiana*, *Rehderia georgiana*, *Bathyaurinia garciai*, and *Bathyaurinia aguayoi* are known from the eastern United States. In stark contrast, the Gulf of Mexico houses three of these five taxa along with twenty others, with endemics such as *Aurinia kieneri*, *Caricellopsis contoyensis*, and *Garciavoluta mcginnorum* being among the largest North American gastropods.

Endemic Mollusks of the Gulf Escarpments, Petroleum Seeps, and Brine Pools

The western half of the Gulf of Mexico is dominated by a single large deep basin that is referred to as the **Sigsbee Deep**. At 5,203 m (17,070 feet) at its deepest, the Sigsbee Deep can be considered to be a true abyssal system (Abyssal Zone) and its flat sea floor qualifies as a true Abyssal Plain (Figure 0.2, Introduction). To the north of this Abyssal Plain, and running along the Louisiana and Texas coasts, is the immense **Sigsbee Escarpment**; a cliff-like feature at the edge of the continental shelf that abruptly drops thousands of meters into the Sigsbee Deep. The base of the escarpment is composed of a rough-textured zone of raised knolls and small depressions, the result of once-buried salt diapirs (salt domes) rising to the sea floor surface. Where these salt diapirs are exposed, they dissolve and form large pools and small lakes of super-dense hypersaline water, many times denser than the normal sea water around them. These **hypersaline brine pools** are often associated with seeps of raw petroleum and methane gas that are squeezed out of the sediments at the top of the diapir by the rising body of salt. The organic compounds associated with the petroleum, along with the dissolved methane gas, act as food resources for methane-feeding bacteria that occur in dense concentrations around the brine pools.

The **methanophilic bacteria** form the base of an entire complex abyssal ecosystem, unlike any seen elsewhere in the western Atlantic. Immense numbers of large mussels live on the "beach" areas surrounding these brine pools, attached to the substrate by strong byssal threads and filtering the bacteria out of the sea water as their food source. Three species of these giant mussels are presently known from the pools, including *Bathymodiolus brooksi*, *Bathymodiolus heckerae*, and *Gigantidas childressi* (see Gustafson *et al.*, 1998) (Figure 2.9). Living on the shells of the giant bacteria-feeding mussels is the bizarre neritoidean gastropod, *Thalassonerita naticoides* (Figure 2.9; Family Phenacolepadidae), which feeds on the bacterial slimes that cover the backs of the living mussels (Clarke, 1989). Living within or near the clumps of mussels are other unusual gastropods such as the gazid *Gaza superba* (a wide-ranging deep water species), the extremely turreted volute *Clenchina worki* (Figure 2.14), the cancellariid *Cancellaria rosewateri* (Figure 2.9), and the raphitomid turroidean *Mioawateria extensa judithae*. An entire methane and petroleum seep-loving bivalve fauna also occurs along with these odd gastropods, including the giant limid *Acesta bullisi* (Figure 2.9), the kelliellids *Vesicomya cordata*, *Vesicomya venusta*, and *Calyptogena ponderosa*, and the thyasirids *Spinaxinus emicatus* and *Thyasira oleophila*. This assemblage of odd petroleophilic mollusks develops wherever brine pools and oil seeps occur anywhere along the escarpments that encircle the Sigsbee Abyssal Plain and creates a special marine ecosystem that is unique to the Gulf of Mexico.

TABLE 3. Important Mollusks that are restricted to the Gulf of Mexico.

Gastropoda

Calliostomatidae
Calliostoma fascinans Schwengel and McGinty, 1942
Calliostoma dentatum Quinn, 1992
Calliostoma springeri Clench and Turner, 1960

Liotiidae
Cyclostrema tortuganum (Dall, 1927)

Neritidae
Nerita (Theliostyla) lindae Petuch, 1988 (also southeastern Florida)

Phenacolepadidae
Thalassonerita naticoides (Clarke, 1989) (bathyal petroleum seeps, on mytilids)

Littorinidae
Littoraria irrorata sayi (Philippi, 1846)

Turritellidae
Vermicularia fargoi Olsson, 1951

Crepidulidae
Crepidula maculosa Conrad, 1846
Crepidula ustulatulina Collin, 2002
Ianacus atrasolea (Collin, 2002)

Naticidae
Neverita (Glossaulax) delessertiana (Recluz, 1843)
Neverita (Glossaulax) duplicata campechiensis (Recluz, 1843)

Muricidae
Calotrpohon andrewsi E. Vokes, 1976
Calotrophon ostrearum (Conrad, 1846)
Paziella nuttingi (Dall, 1896)
Stramonita canaliculata (Gray, 1839)
Stramonita canaliculata form *haysae* Clench, 1927

Fasciolariidae
Aristofusus helenae (Bartsch, 1939) (erroneously reported from South America, but those refer to other species)
Fasciolaria bullisi Lyons, 1972
Heilprinia timessa (Dall, 1889) (also southeastern Florida)

Pisaniidae
Solenosteira cancellaria (Conrad, 1846) (Suwannean and Texan Subprovinces)

Busyconidae
Fulguropsis plagosus (Conrad, 1862)

Columbellidae
Columbella rusticoides Heilprin, 1887 (also NW Cuba)

Volutidae
Aurinia dubia (Broderip, 1827)
Aurinia kieneri (Clench, 1946)
Clenchina dohrni (Sowerby III, 1903) (Floridian and Suwannean Subprovinces)
Clenchina robusta Dall, 1889
Clenchina worki Coltro, 1998

Cancellariidae
Cancellaria rosewateri Petit, 1983
Ventrilia tenerum (Philippi, 1848)

Conidae
Conasprelloides cancellatus (Hwass, 1792)
Dalliconus armiger (Crosse, 1858)
Dalliconus rainesae (McGinty, 1953)

Dalliconus sauros (Garcia, 2006) (Texan and Yucatanean Subprovinces)
Gradiconus largillierti (Kiener, 1847)

Terebridae

Neoterebra arcas (Abbott, 1954)

Turridae

Polystira tellea (Dall, 1889)

Drilliidae

Cerodrillia clappi Bartsch and Rehder, 1939
Cerodrillia perryae Bartsch and Rehder, 1939
Cerodrillia schroederi Bartsch and Rehder, 1939
Fenimorea falloni Garcia, 2023

Raphitomidae

Mioawateria extensa judithae (Clarke, 1989) (bathyal petroleum seeps)

Mangeliidae

Brachycythara barbarae Lyons, 1972

Bivalvia

Pectinidae

Argopecten irradians taylorae (Petuch, 1987) (Floridian and Suwannean Subprovinces)
Euvola marshallae Petuch and Myers, 2014

Limidae

Acesta bullisi (H. Vokes, 1963)

Mytilidae

Bathymodiolus brooksi Gustafson, Turner, Lutz, and Vrijenhoek, 1998 (bathyal petroleum seeps)
Bathymodiolus heckerae Gustafson, Turner, Lutz, and Vrijenhoek, 1998 (bathyal petroleum seeps)
Gigantidas childressi (Gustafson, Turner, Lutz, and Vrijenhoek, 1998) (bathyal petroleum seeps)

Kelliellidae

Calyptogena ponderosa (Boss, 1968)
Vesicomya cordata (Boss, 1968)
Vesicomya venusta Dall, 1886 (bathyal petroleum seeps)

Thyasiridae

Spinaxinus emicatus Oliver, Rodrigues, Carney, and Duperron, 2013 (bathyal petroleum seeps)
Thyasira oleophila Clarke, 1989 (bathyal petroleum seeps)

Cyrenidae

Polymesoda floridana (Conrad, 1846)

Cardiidae

Dinocardium vanhyningi Clench and Smith, 1944

Veneridae

Anomalocardia auberiana (d'Orbigny, 1853)
Anomalocardia cuneimeris (Conrad, 1846)

ICONOGRAPHY OF WIDESPREAD GULF OF MEXICO ENDEMIC SPECIES

Figure 2.3 Mississippi Sea ancestors of the modern Gulf of Mexico endemic fauna.
A, B= *Spinifulgur spiniger* (Conrad, 1848), length 39 mm, early Oligocene (Busyconidae); **C, D=** *Atraktus demissus* (Conrad, 1848), length 36 mm, early Oligocene (Volutidae); **E=** *Atraktus texanus* (Gabb, 1860), length 32 mm, early Oligocene (Volutidae); **F=** *Reticulacella reticulata* (Aldrich, 1885), length 30 mm, Oligocene (Volutidae); **G, H=** *Caricella bolaris* (Conrad, 1833), length 25 mm, late Eocene (Volutidae); **I, J=** *Caricella polita* (Conrad, 1854), length 33 mm, late Eocene (Volutidae). These Mississippi Sea genera are the direct ancestors of the living scaphelline volute and busycon whelk faunas of the modern Gulf of Mexico.

Figure 2.4 Widespread Neritic Zone Gastropods of the Gulf of Mexico.
A= *Solenosteira cancellaria* (Conrad, 1846), length 24 mm. Low tide, Galveston, Texas; **B, C=** *Calliostoma springeri* Clench and Turner, 1960, diameter 28 mm, deep water off the Dry Tortugas; **D, E=** *Crepidula ustulatulina* Collin, 2002, length 15 mm, found on *Trochomodulus* shells in Florida Bay, Florida; **F=** *Calotrophon ostrearum* (Conrad, 1846), 23 mm, in Turtle Grass near mangroves, Little Torch Key, Florida Keys; **G=** *Calotrophon andrewsi* E. Vokes, 1976, length 21 mm, 50 m depth west of Naples, Florida; **H=** *Columbella rusticoides* Heilprin, 1887, length 18 mm, in a Turtle Grass bed off Joe Kemp Key, Florida Bay; **I, J=** *Nerita (Theliostyla) lindae* Petuch, 1988, width 15 mm, buried in sand under rocks, Lake Worth, Florida. Also found along Mississippi and Louisiana; this tweed-patterned nerite is frequently overlooked because of its cryptic lifestyle (see Petuch, 1988; Petuch and Myers. 2014).

Figure 2.5 ***Stramonita canaliculata*** **and Variety** ***haysae*****, endemic to the Gulf of Mexico.**
A, B= *Stramonita canaliculata* (Gray, 1839), length 63 mm, Port Aransas, Texas; **C, D=** *Stramonita canaliculata* (Gray, 1939), length 66 mm, Galveston, Texas; **E, F=** *Stramonita canaliculata* form *haysae* Clench, 1927, length 72 mm, Mobile Bay, Alabama.

Figure 2.6 *Fulguropsis plagosus* from the northern and eastern Gulf of Mexico.
A, B= *Fulguropsis plagosus* (Conrad, 1862), length 72 mm, off the Dry Tortugas; **C, D**= *Fulguropsis plagosus* (Conrad, 1862), length 104 mm, Alligator Point, Florida. This offshore, deeper water species ranges from the Mississippi River Delta to the Dry Tortugas, but it is most commonly encountered off the Florida Panhandle and Alabama.

Figure 2.7 ***Dinocardium vanhyningi*** **from the entire Gulf of Mexico.**
A, B= *Dinocardium vanhyningi* Clench and Smith, 1944, dorsal and side views, length 135 mm; **C=** *Dinocardium vanhyningi* Clench and Smith, 1944, length 61 mm. Both specimens from Camp Lulu Key, Ten Thousand Islands, Collier County, Florida. This Gulf of Mexico endemic species is often incorrectly considered to be a form or subspecies of the southeastern United States *Dinocardium robustum* (Lightfoot, 1786), which ranges from North Carolina to Cape Canaveral, Florida. That Georgian Subprovince endemic differs from the Gulf *vanhyningi* in having fewer and coarser ribs and in having a distinctly rounder shell outline. There are no *Dinocardium* species of any kind in the Florida Keys, and the carbonate environments of the Floridian Subprovince act as ecological barriers between the two sibling species. When comparing the two shells, *vanhyningi* can be seen to always be much more elongated, with more numerous and smoother ribs.

Figure 2.8 Deep Neritic Mollusks of the Gulf of Mexico.
A= *Fasciolaria bullisi* Lyons, 1972, length 159 mm, eastern side of the Yucatan Peninsula; **B, C**= *Fasciolaria bullisi* Lyons, 1972, length 180.0 mm, eastern side of the Yucatan Peninsula; **D, E**= *Ventrilia tenerum* (Philippi, 1848), length 24 mm, off the Dry Tortugas; **F**= *Polystira tellea* (Dall, 1889), length 66 mm, 250 m depth off the Dry Tortugas; **G**= *Fenimorea falloni* Garcia, 2023, length 53.6 mm, from 100 m depth off Cabo Catoche, Quintana Roo State, Mexico.

Figure 2.9 Widespread Bathyal Mollusks of the Gulf of Mexico.
A= *Euvola marshallae* Petuch and Myers, 2014, width 46 mm, south of the Dry Tortugas, Florida; **B=** *Bathymodiolus brooksi* Gustafson, Turner, Lutz, and Vrijenhoek, 1998, length 93.8 mm, on a petroleum seep, 229 m depth off Vermilion Parish, Louisiana; **C, D=** *Paziella nuttingi* (Dall, 1986), length 42 mm, off the Dry Tortugas; **E, F=** *Cancellaria rosewateri* Petit, 1983, length 35 mm, in a *Gigantidas* mussel bed, Alaminos Canyon, Sigsbee Escarpment; **G=** *Thalassonerita naticoides* (Clarke, 1989), length 9.4 mm, on the back of a *Gigantidas* mussel shell, from a petroleum seep in the Alaminos Canyon, Sigsbee Escarpment; **H=** *Gigantidas childressi* (Gustafson, Turner, Lutz, and Vrijenhoek, 1998), length 101 mm, on a petroleum seep, Alaminos Canyon, Sigsbee Escarpment; **I=** *Acesta bullisi* (H. Vokes, 1963), length 120 mm, Alaminos Canyon, Sigsbee Escarpment. The living Gulf *Euvola marshallae* has been erroneously referred to the taxon *E. papyracea* (Gabb, 1873) by most previous workers; that name, however, is referable only to a fossil species that is found in the early Miocene beds of the Dominican Republic.

Figure 2.10 ***Aurinia dubia*** **from the Gulf of Mexico Bathyal Zone.**
A, B= *Aurinia dubia* (Broderip, 1827), length 135 mm, 100 m depth west of Sarasota, Florida; **C, D=** *Aurinia dubia* (Broderip, 1827), length 135.0 mm, 150 m depth south of Apalachicola, Florida. This species differs from the similar-appearing *A. kieneri* (Figure 2.11) in being a smaller and more slender shell with a narrower shoulder and in having a more elongated protoconch that is most often dark brown or tan in color. *Aurinia dubia* also lives in much shallower water than does *A. kieneri*, which most often occurs in depths of over 250 m.

Figure 2.11 ***Aurinia kieneri*** **from the Texan and Suwannean Subprovinces.**
A, B= *Aurinia kieneri* (Clench, 1946), length 198 mm, 250 m depth south of Dauphin Island, Alabama; **C, D**= *Aurinia kieneri* (Clench, 1946), holotype, length 197 mm, 270 m depth off the Mississippi River Delta. Similar in appearance to *A. dubia*, but differs in being a much larger, thinner, and more inflated shell with a rounded protoconch that is a pale cream-yellow color.

Figure 2.12 *Clenchina dohrni* from the Floridian Subprovince Bathyal Zone.
A, B= *Clenchina dohrni* (Sowerby III, 1903), length 53 mm, 300 m depth off Key West, Florida Keys; **C=** *Clenchina dohrni* (Sowerby III, 1903), length 62 mm, 300 m depth off Key West, Florida; **D, E=** *Clenchina dohrni* (Sowerby III, 1903), length 48 mm, 300 m depth south of the Dry Tortugas, Florida.

Figure 2.13 ***Clenchina robusta*** **from the Gulf of Mexico Bathyal Zone.**
A, B= *Clenchina robusta* (Dall, 1889), length 80 mm, 400 m depth south of Key West, Florida Keys; **C, D=** *Clenchina robusta* (Dall, 1889), length 50.0 mm, 350 m depth north of the Dry Tortugas; **E, F=** *Clenchina robusta* (Dall, 1889), holotype, length 119 mm, 512 m depth south of Key West, Florida.

Figure 2.14 *Clenchina worki* from the Texan Subprovince Bathyal Zone.
A, B= *Clenchina worki* (Coltro, 1998), length 72 mm, 475 m depth, Alaminos Canyon, Sigsbee Escarpment; **C, D=** *Clenchina worki* (Coltro, 1998), juvenile specimen, length 36.5 mm, 475 m depth, Alaminos Canyon, Sigsbee Escarpment.

Figure 2.15 ***Gradiconus largillierti*** **from the northern, western, and southwestern Gulf of Mexico. A, B=** *Gradiconus largillierti* (Kiener, 1847), length 40 mm, off Panama City, Florida; **C=** *Gradiconus largillierti* (Kiener, 1847), length 43 mm, off Panama City, Florida; **D, E=** *Gradiconus largillierti* (Kiener, 1847), length 41 mm, off Contoy Island, Yucatan Peninsula, Mexico.

Figure 2.16 Cone Shells of the Gulf of Mexico.
A, B= *Conasprelloides cancellatus* (Hwass, 1792), length 35 mm, off Grande Isle, Louisiana; **C, D=** *Dalliconus rainesae* (McGinty, 1953), length 17 mm, Campeche Bank, Yucatan Peninsula, Mexico; **E, F=** *Dalliconus armiger* (Crosse, 1858), length 35 mm, off Grande Isle, Louisiana; **G, H=** *Dalliconus rainesae* (McGinty, 1953), length 30 mm, north of the Dry Tortugas, Florida.

Figure 2.17 *Dalliconus sauros* from the Texan and Yucatanean Subprovinces.
A, B= *Dalliconus sauros* (Garcia, 2006), holotype, length 29.5 mm, trawled from 60 m depth, 58 km SE of Port Aransas, Texas; **C, D**= *Dalliconus sauros* (Garcia, 2006), length 34.2 mm, trawled from 60 m depth off Cabo Catoche, Quintana Roo State, Mexico; **E, F**= *Dalliconus sauros* (Garcia, 2006), length 15.3 mm, 60 m depth off Cabo Catoche, Quintana Roo State, Mexico; **G, H**= *Dalliconus sauros* (Garcia, 2006), length 37 mm, 100 depth east of Veracruz, Mexico.

Plate 2.18 *Neverita* (*Glossaulax*) species (Naticidae) that are endemic to the Gulf of Mexico.
A, B= *Neverita (Glossaulax) delessertiana* (Recluz, 1843), width 33 mm, sand flats off Rabbit Key, Ten Thousand Islands, Monroe County, Florida; **C**= base of the same specimen of *delessertiana*, showing the small umbilical callus that covers only half the umbilical area; **D, E**= *Neverita (Glossaulax) duplicata campechiensis* (Recluz, 1843), width 54 mm, Kice Island, Ten Thousand Islands, Collier County, Florida; **F**= base of the same specimen of *campechiensis*, showing the large umbilical callus that covers three-quarters of the umbilicus, leaving only a small, deep pit. The nominate subspecies, *N. duplicata duplicata* (Say, 1822) from the eastern coast of the United States, has a proportionally much larger umbilical callus that covers the entire umbilical opening. The Florida Keys coral reef areas and carbonate environments act as an ecological barrier between *duplicata* and *campechiensis* and the two subspecies are genetically isolated from each other. Compare these two *Neverita* species with the Texan endemic, *Neverita (Glossaulax) duplicata texasiana*, shown in Figure 5.15 in Chapter 5.

Gaza superba (Dall, 1881), width 32 mm, from 1200 m depth in the Sigsbee Deep. This widespread deep water species ranges throughout the Caribbean Basin and is common within the bathyal and abyssal areas of the Gulf of Mexico. When disturbed, the animal violently undulates its elongated foot and actually swims away from potential danger.

CHAPTER 3. Endemic Mollusks of the Floridian Subprovince

Within the Gulf of Mexico region, the Floridian Subprovince extends along southern Florida, from Palm Beach to Biscayne Bay, and then westward across the entire **Florida Keys** as far as the **Dry Tortugas** (Petuch, 2013; Petuch and Myers, 2014; Petuch and Berschauer, 2021a). The subprovincial boundaries are defined, essentially, by the configuration of the **Florida Coral Reef Tract**, the largest contiguous coral structure found in the continental United States (Figure 3.1). Besides living coral reefs, immense areas of fine carbonate sediments have been deposited and these cover large swaths of the broad, shallow **Florida Platform**. The carbonate sea floors in this area support extensive beds of Turtle Grass (*Thalassia testudinum*) and other sea grasses, as well as large bioherms of *Spheciospongia* and *Ircinia* sponges and delicate branching corals such as *Porites diverticulata* and *Porites furcata*. Throughout the subprovince, these three main habitat types blur together into an ecological continuum of biotopes that, altogether, supports the single richest molluscan fauna found anywhere in the Gulf of Mexico.

The living coral reefs, with their raised topography and diverse microhabitats, house the most species-rich malacofauna found within the subprovince and support a large number of endemic species. Primary among these locally-restricted taxa is a large radiation of Flamingo Tongue Shells (genus *Cyphoma* of the Family Ovulidae), the largest and richest in the entire western Atlantic (Petuch and Myers, 2014: 95-98). These interesting shells (some shown here in Figure 3.4), with their individually-marked and species-specific colored mantles, reside on living gorgonian octocorals, such as members of the genera *Pseudopterogorgia*, *Gorgonia*, *Plexaurella*, and *Plexaura*, and graze on the extended polyps. Three widespread species, *Cyphoma gibbosum*, *C. mcgintyi*, and *C. signatum* are most commonly-encountered on the Floridian gorgonian forests and these live along with the rarer endemic Florida Keys *Cyphoma rhomba* and *C. sedlaki*. A sixth rare *Cyphoma* species, *C. alleneae*, also lives on shallow reef areas along the Florida Reef Tract, but unlike its congeners, it feeds on living *Porites* corals instead of gorgonians. The deep subtidal areas (20-100 m depths) along the seaward side of the reef tracts support large “forests” of the antipatharian octocorals *Antipathes* and *Cirrhipathes* and these house three species of *Cyphoma*; the widespread western Atlantic *C. aureocinctum* and *C. intermedia* and the rare and poorly-known Florida endemic, *C. gibbulum*. As presently understood, the Floridian Subprovince *Cyphoma* fauna contains nine species, making it the most species-rich in the entire Atlantic and Eastern Pacific Oceans. The Floridian Subprovince endemic ovulids are shown here in Figure 3.4.

Another gastropod family that has evolved many distinctive endemic species on the coral reefs of the Floridian Subprovince is the Muricidae, the Murex Shells. These predatory mollusks mostly feed by drilling through the shells of other mollusks, using their hard radular teeth and buccal acids to both create a hole in the shell and also to access the exposed flesh of their prey. Some specialized muricids, such as the *Coralliophila* and *Babelomurex* Coral Shells, feed on the mucus exuded by corals and also on their body

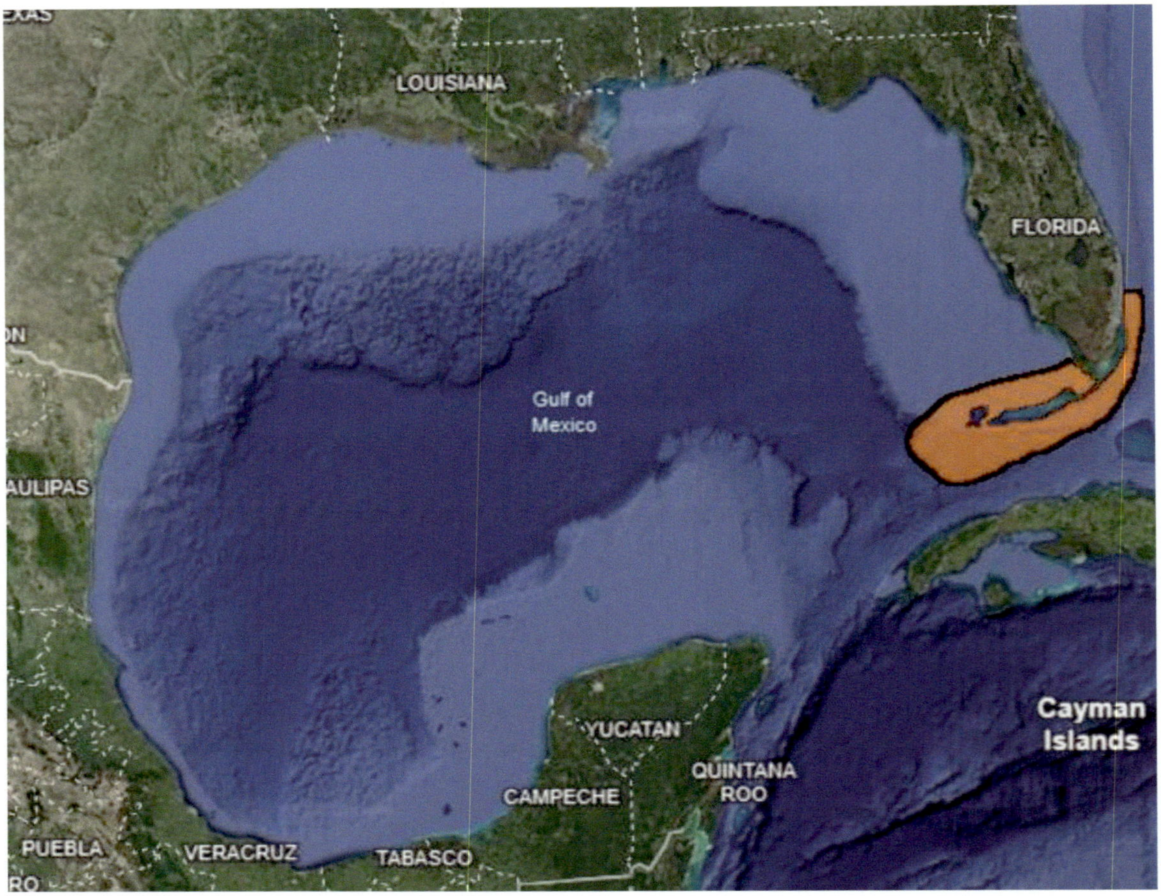

Figure 3.1 Map showing the geographical extent of the Floridian Subprovince (orange shading) of the Carolinian Molluscan Province, which contains the entire **Florida Coral Reef Tract** and the Dry Tortugas reef complexes.

fluids. To date, nine species of muricids have been found to be endemic to the coral reefs of southern Florida and the Florida Keys Reef Tract, and these encompass six genera; *Favartia*, *Dermomurex*, *Murexiella*, *Pygmaepterys*, *Vokesimurex*, and *Coralliophila*. Some, such as the small species *Dermomurex pacei* (Figure 3.8 A, B) live inside eroded cavities within the holdfasts of large gorgonians, where they feed on small encrusting bivalves such as members of the genera *Dimya* and *Plicatula*. Other Floridian Subprovince muricid reef endemics, such as *Coralliophila kalafuti* (Figure 3.8 C, D), live associated with masses of *Porites porites* corals on the deep reefs at the base of the main reef tracts. Other endemic muricids, such as *Favartia pacei* (Figure 3.8 E, F) and *Murexiella caitlinae* (Figure 3.8 G, H) live in association with beds of their prey bivalve, the Yellow Mussel *Brachidontes modiolus*. The limestone platforms and fossil coral reefs that are exposed along the intertidal zone of the Florida Keys, adjacent to the shallow reefs, also house a number of prominent endemic gastropods. Of particular interest is the dwarf melongenid *Melongena (Rexmela) bicolor* and its forms *minor* and *estephomenos* (Figures 3.12 and 3.13) and the endemic areneid *Arene vanhyningi* (Figure 3.4).

The Turtle Grass beds of the carbonate banks of the Floridian Subprovince contain an equally-rich molluscan fauna as does the coral reef tract. Many of the species found among the shallow sea grass meadows are restricted to the Florida Keys and their presence underscores the uniqueness of the local *Thalassia*-associated malacofauna. Some of these sea grass endemics include the calliostomatid *Calliostoma adelae* (Figure 3.4 A, B), the modulid *Trochomodulus calusa* (Figure 3.6 C-E), the turbinid *Lithopoma americana* (Figure 3.6 F, G), the small banded bullid *Bulla frankovichi* (Figure 3.7, G-I), the haminoeid *Haminoea taylorae* (Figure 3.6, H-J), the columbellid *Falsuszafrona taylorae* (Figure 3.5 E, F), and the small cone shell *Jaspidiconus pealii* (Figure 3.21 G, H). The Turtle Grass (*Thalassia*) beds also house a special Gulf endemic muricid, the delicate, elongated *Vokesimurex marcoensis* (Figure 3.5 C, D). Although named for Marco Island along southwestern Florida, this previously-overlooked and newly-rediscovered species is restricted to the sea grass and sponge beds in the Lower Keys, ranging from Big Pine Key westward to the Dry Tortugas; and is not found on Marco Island. The Florida Keys Whelk, *Fulguropsis keysensis* (Figure 3.11), a coarsely-sculptured small busyconid, also lives in the sea grass beds, where it feeds upon small bivalves that live nestled among the *Thalassia* rhizomes. Although primarily a Keys organism, *F. keysensis* is now known to occur in the adjacent Ten Thousand Islands of south western Florida (see Chapter 4), where it is sympatric with the closely-related *F. pyruloides*. The endemic nassariids, *Uzita swearingeni* and *Uzita websteri* (both Figure 3.5), and the endemic cone, *Jaspidiconus acutimarginatus* (Figure 3.21; see Eason, 2018), are also found in open sand patches between individual sea grass beds. All of these endemic species and the ecological structure of their resident Florida Keys Turtle Grass beds are discussed in detail in Petuch and Myers (2014).

In the Keys, large bioherms composed of the sponges *Spheciospongia* and *Ircinia* and the Finger Coral (*Porites furcata*) are found in close association with the Turtle Grass beds, and these two biotopes often intermingle at many localities along the Lower Florida Keys and Dry Tortugas. One of the more conspicuous gastropods of the sea grass beds is the large endemic cone shell, *Gradiconus burryae* (Figure 3.17), which feeds on polychaete worms living among the *Thalassia* rhizomes. As seen here, Burry's Cone is quite variable in color, being most often a solid deep chestnut brown with darker thin bands. Some individuals are much paler, being white with scattered zig-zag patterned brown flammules. In the adjacent sponge and coral bioherms, another endemic Keys *Gradiconus* species is now known to occur; the small *G. mazzolii* (Figure 3.18, A-G). This diminutive cone has a highly-restricted range, being found only on shallow water sponge bioherm areas from Big Pine Key west to the Dry Tortugas. Although similar in appearance to the larger, Turtle Grass-dwelling *Gradiconus burryae*, the sponge-associated *G. mazzolii* has a more slender shell, a much higher and more protracted spire, and a different-colored animal; *G. burryae* has an orange animal while *G. mazzolii* has a bright yellow animal (shown here in Figure 3.19).

Figure 3.2 View of the shoreline on Missouri Key, Middle Florida Keys, Monroe County, Florida; looking south to the Missouri-Ohio Channel and Ohio Key. This shoreline, and the areas just offshore, are composed of an exposed outcrop of the late Pleistocene Key Largo Formation and house a large number of endemic Floridian Subprovince mollusks. Some of these, which are found on the shallow reef and sea grass beds just off the island, include *Murexiella caitlinae*, *Favartia pacei*, *Gradiconus burryae*, and *Jaspidiconus pealii*. The dense growth of gorgonian octocorals that grows right offshore (the dark color band) supports an endemic Keys ovulid fauna, including *Cyphoma alleneae*, *Cyphoma sedlaki*, and *Cyphoma rhomba*. These co-exist with the widespread *Cyphoma gibbosum*, *Cyphoma mcgintyi*, *Cyphoma signatum*, *Cyphoma intermedium*, and *Simnialena uniplicata* to produce the richest ovulid fauna found at any single locality within the tropical western Atlantic.

Figure 3.3 View of Missouri Key, Middle Florida Keys, Monroe County, Florida at low tide, showing the exposed coral limestone shoreline and an approaching severe thunderstorm with its accompanying tornadic water spout. The tide pools that form on this exposed fossil coral reef (Key Largo Formation; late Pleistocene) house a rich fauna of mollusks, including the Floridian Subprovince endemic gastropods *Calliostoma adelae*, *Lithopoma americana*, *Murexiella caitlinae*, and *Melongena (Rexmela) bicolor*.

On the offshore carbonate environments, another closely-related endemic cone shell, *Gradiconus patglicksteinae* (Figure 3.20), occurs in association with living corals. Found throughout the entire reef tract, this elongated species has the most variably-colored shell of any Florida *Gradiconus* species, varying from almost pure white to bright pink and deep orange-red. First collected in deep water along the coral reefs off Palm Beach, the holotype specimen was a bright orange and pink shell and was thought to be typical of all

G. patglicksteinae. Subsequent collecting near living coral reefs farther south off Plantation Key brought to light similarly-shaped specimens, but with white shells and scattered orange patches. These were later named *Gradiconus tranthami* in 1995 by the senior author and were thought to be a distinct reef-dwelling species. In 2011, bright pink specimens of this species were found on the coral reefs around the Dry Tortugas and were named *G. antoni* by Cargile and *G. tortuganus* by Petuch and Sargent (now a *nomen nudum*, as the publisher failed to publish the description in the original publication). Interestingly enough, bright orange specimens, exactly like *G. patglicksteinae*, and white specimens, similar to *G. tranthami*, have been collected along with the pink *antoni* form on the Dry Tortugas. This clearly shows that all three color forms represented a single variable coral reef species (all three forms are shown here in Figure 3.20). The nomenclatural clarification of *Gradiconus patglicksteinae* demonstrates that the genus *Gradiconus* has recently undergone an evolutionary radiation, producing offshoot sibling species that have moved into, and have become ecologically-restricted to, three main types of environments; Turtle Grass beds, sponge bioherms, and living coral reefs. This species radiation can be traced back to an ancestor found in the Middle Pleistocene (Calabrian Age) fossil beds of southern Florida; *Gradiconus loxahatcheensis* Petuch, 1994 of the Holey Land Member of the Bermont Formation. This ancestral cone gave rise to a complex that includes these three species and also *G. floridanus* and *G. largillierti* (see Petuch, Drolshagen, and Herndl, 2015; Petuch and Berschauer, 2021a).

Living along with *Gradiconus patglicksteinae* in carbonate sand areas between living coral heads, are several predatory volutoidean gastropods that are endemic to the Floridian Subprovince. Primary among these is the newly-described dwarf scaphelline volute, *Scaphella glicksteinorum* (Figure 3.15). Unlike the other scaphelline volutes of the Gulf of Mexico, this dwarf species, the smallest member of the *Scaphella* (*s.s.*) complex, lives in sand pockets on the main reef platform and has been collected in water as shallow as 3 m depth. Here, it feeds on the endemic Floridian Subprovince olive shell, *Americoliva bollingi* (Figure 3.16) in a similar fashion to its larger relative, *Scaphella junonia*, which hunts and devours the olive shell *Americoliva sayana sarasotaensis* along southwestern Florida. In slightly deeper depths of 10-20 m off the Dry Tortugas and areas along the western Florida Shelf, the larger, inflated *Scaphella elizabethae* (Figure 3.14) replaces its dwarf relative in similar carbonate ecosystems. These distinctive endemic volutoideans occur together with other important coral reef-associated endemic gastropods such as the reef mitrid *Dibaphimitra florida* (Figure 3.7), the brightly-colored *Fasciolaria bittneri* (Figure 3.27; see Petuch and Berschauer, 2023a), and the cancellariid *Cancellaria adelae* (Figure 3.7). The deeper areas off the Dry Tortugas and the Lower Keys are now known to house a very rich molluscan fauna that has a large component of endemic species. Some of these include the vermetid *Cayo margarita*, the muricid *Murexiella kalafuti* (Figure 3.9), and the fasciolariid *Leucozonia jacarusoi* (Figure 3.5).

The extensive areas of carbonate sand and shell rubble on the offshore terraces also support a rich and highly endemic molluscan fauna. On the shallower upper terraces, numerous vermivorous cone shells dominate the sea floor, including distinctive species such as *Jaspidiconus fluviamaris* (Figure 3.21), *J. vanhyningi* (Figure 3.21), *Tuckericonus caribaeus* (Figure 3.22), *T. flamingo* (Figure 3.23), *Kellyconus binghamae* (Figure 3.24), and *Dauciconus glicksteini* (Figure 3.25). This large fauna of conids occurs together with other sand-dwelling endemic species such as the olivids *Americoliva matchetti* (Figure 3.28 A-D) and *A. recourti* (Figure 3.28 E-H) and the marginellid *Prunum frumari* (Figure 3.26 A, B). The deepest terraces bordering the Dry Tortugas and Florida Keys contain one of the richest faunas of scaphelline volutes found in the entire Gulf of Mexico, with widespread Gulf species such as *Clenchina dohrni*, *Clenchina gouldiana*, and *Clenchina robusta* (see Chapter 2) occurring with Floridian Subprovince endemics such as *Clenchina florida* (Figure 3.30), *Clenchina atlantis* (Figure 3.29), *Clenchina rundoi* (Figure 3.31), *Clenchina carrieae* (Figure 3.34), and *Rehderia schmitti* (Figure 3.33). These rarely-studied and poorly-sampled deep terrace ecosystems have been shown to house other odd endemic gastropods, such as the colubrariid *Bartschia frumari* (Figure 3.26).

Endemic Mollusks of the Deep Reefs

The entire Florida Reef Tract plummets abruptly into the Straits of Florida after only a few kilometers distance from shore. The steep slope of the fore-reef area drops down, at an acute angle, to a series of narrow terraces, one after the other in a step-like fashion. These flat terraces represent ancient shorelines which formed during times of lower sea levels in the Pliocene and Pleistocene Epochs. Because of technical and logistical difficulties in sampling these steep areas, the slopes between the individual terraces have rarely been visited or studied, but the little information that is available has shown that these deep fore-reef areas contain a molluscan fauna with a high level of biodiversity and endemism. Some areas of the Florida deep reef system are unique in the Gulf of Mexico, supporting highly unusual ecosystems like those seen on the **Pulley Ridge** near the Dry Tortugas. Here, in the **Mesophotic Zone** at 60-100 meters depths, the deepest coral reefs in the Gulf of Mexico are known to exist and are composed almost entirely of the Lettuce Corals *Agaricia* and *Leptoceris*. In the perpetual twighlight of this deep offshore area, the corals take on an intense purple color, the result of containing commensal dinoflagellates (zooxanthellae) in their tissues. The purple pigments in these microscopic algae have evolved for utilizing the little sunlight that does penetrate to these lowermost Neritic Zone depths. Other areas in the deep reef system, such as those on the 30 m terrace off the Middle and Lower Florida Keys, shelter extensive forests of gorgonian and antipatharian octocorals.

The fragile corals of the deep reef areas along extreme southeastern Florida and the Florida Keys shelter a large number of small and rarely-seen endemic gastropods, some of which include the calliostomatid *Calliostoma frumari*, the triviid *Pusula juyingae* (Figure 3.6 A, B), muricids such as *Dermomurex (Trialatella) glicksteini* (Figure 3.9 C-E), *Pygmaepterys richardbinghami* (Figure 3.9 I, J), *Timbellus guesti* (Figure 3.9 F), and

Favartia goldbergi (Figure 3.9 G, H), and cancellariids like *Tritonoharpa janowskyi* (Figure 3.26 J). In the microhabitats found among the coral rubble and eroded coral heads, these small gastropods are the apex predators on infaunal invertebrates and encrusting bivalves. The near-vertical walls of the deep fore-reef area support dense growths of delicate *Agaricia* Lettuce Coral and stylasterine hydrocorals and shelter a number of delicate sessile bivalves that are unique to the Florida Reef Tract. Some of these include the rare and newly-discovered arcid, *Arca rachelcarsonae* (Figure 3.26 E, F, G) and the ornate and fragile chamid *Amphichama inezae*. Of special interest on these deep coral walls is the beautiful and colorful endemic Florida Reef Tract pectinid *Caribachlamys mildredae* (Figure 3.26 H, I), whose behavior of attaching itself with byssal threads under slabs of coral makes it almost impossible to collect. This cryptic sessile scallop also serves as the host animal for the endemic ectoparasitic amathinid gastropod, *Amathina pacei* (Figure 3.7 C, D). This small, flattened shell typically attaches itself to the back of the *Caribachlamys* in the area near the scallop's auricles. From this attachment spot, the amathinid extends its long proboscis into the commissure of the valves and pierces the scallop's exposed mantle tissue, sucking the body fluids like a mosquito. Some rare deep reef Caribbean taxa that also occur in the Gulf are shown here in Figure 3.35.

TABLE 4. Important Mollusks Endemic to the Floridian Subprovince.

Gastropoda

Calliostomatidae

Calliostoma adelae Schwengel, 1951
Calliostoma frumari Garcia, 2007

Turbinidae

Lithopoma americana (Gmelin, 1791)

Areneidae

Arene vanhyningi Rehder, 1943

Ovulidae

Cyphoma alleneae Cate, 1973
Cyphoma gibbulum Cate, 1973
Cyphoma rhomba Cate, 1978
Cyphoma sedlaki Cate, 1976

Triviidae

Pusula juyingae Petuch and Sargent, 2011

Modulidae

Trochomodulus calusa (Petuch, 1988)

Vermetidae

Cayo margarita Bieler, Collins, Golding, and Rawlings, 2023

Muricidae

Coralliophila kalafuti (Petuch, 1987)
Dermomurex pacei Petuch, 1988
Dermomurex (Trialatella) glicksteini Petuch, 1987
Favartia goldbergi Petuch and Sargent, 2011
Favartia pacei Petuch, 1988
Murexiella caitlinae Petuch and Myers, 2014
Murexiella kalafuti Petuch, 1987
Pygmaepterys richardbinghami (Petuch, 1987)
Timbellus guesti (Harasewych & Jensen, 1979)

Vokesimurex marcoensis (Sowerby III, 1900)
Vokesimurex rubidus delicatus (M. Smith, 1940)

Fasciolariidae

Cinctura tortugana (Hollister, 1957)
Fasciolaria bittneri Petuch and Berschauer, 2023
Leucozonia jacarusoi Petuch, 1987

Colubrariidae

Bartschia frumari Garcia, 2008

Melongenidae

Melongena (Rexmela) bicolor (Say, 1827)
Melongena (Rexmela) bicolor form *estephomenos* (Melvill, 1881)
Melongena (Rexmela) bicolor form *minor* (Sowerby III, 1879)

Busyconidae

Fulguropsis keysensis Petuch, 2013 (also Ten Thousand Islands)

Nassariidae

Uzita swearingeni (Petuch and Myers, 2014) (also Ten Thousand Islands)
Uzita websteri (Petuch and Myers, 2014)

Columbellidae

Falsuszafrona taylorae (Petuch, 1987)

Pisaniidae

Ameranna florida (Garcia, 2008)

Volutidae

Clenchina atlantis (Clench, 1946)
Clenchina carrieae Petuch and Berschauer, 2023
Clenchina florida (Clench and Aguayo, 1940)
Clenchina rundoi Petuch and Berschauer, 2023
Rehderia georgiana (Clench, 1946)
Rehderia schmitti (Bartsch, 1931)
Scaphella elizabethae Petuch and Sargent, 2011
Scaphella glicksteinorum Petuch and Berschauer, 2023

Mitridae

Dibaphimitra florida (Gould, 1856)

Marginellidae

Prunum frumari Petuch and Sargent, 2011

Olividae

Americoliva bollingi (Clench, 1943)
Americoliva matchetti Petuch and Myers, 2014
Americoliva recourti Petuch and Myers, 2014

Cancellariidae

Cancellaria adelae Pilsbry, 1940
Tritonoharpa janowskyi Petuch and Sargent, 2011

Conidae

Dauciconus glicksteini (Petuch, 1987)
Gradiconus burryae (Clench, 1942)
Gradiconus mazzolii Petuch and Sargent, 2011
Gradiconus patglicksteinae (Petuch, 1987)
Gradiconus patglicksteinae form *antoni* Cargile, 2011
Gradiconus patglicksteinae form *tranthami* Petuch, 1995
Jaspidiconus acutimarginatus (Sowerby II, 1866)
Jaspidiconus fluviamaris Petuch and Sargent, 2011
Jaspidiconus pealii (Green, 1830)
Jaspidiconus vanhyningi (Rehder, 1944)

Kellyconus binghamae (Petuch, 1987)
Tuckericonus caribaeus (Clench, 1942)
Tuckericonus flamingo (Petuch, 1980)

Clathurellidae
Corinnaeturris humilis Garcia, 2024 (possibly endemic to the Dry Tortugas)

Bullidae
Bulla frankovichi Petuch and Sargent, 2011

Haminoeidae
Haminoea taylorae Petuch, 1987

Amathinidae
Amathina pacei (Petuch, 1987)

Bivalvia

Arcidae
Arca rachelcarsonae Petuch and Myers, 2014

Pectinidae
Caribachlamys mildredae (Bayer, 1941)

Chamidae
Amphichama inezae (Bayer, 1943)

Amphichama inezae (Bayer, 1943), a rarely-seen endemic bivalve from the deep reefs and terraces of the Floridian Subprovince.

ICONOGRAPHY OF FLORIDIAN SUBPROVINCE ENDEMIC SPECIES

Figure 3.4 Floridian Subprovince Endemic Species.
A, B= *Calliostoma adelae* Schwengel, 1951, height 17.4 mm, Middle Torch Key, Florida Keys; **C, D**= *Cyphoma sedlaki* Cate, 1876, length 13.5 mm, off Sombrero Light, Florida Keys; **E, F**= *Cyphoma rhomba* Cate, 1978, length 21 mm, Pickles Reef, Florida Keys; **G, H**= *Cyphoma gibbulum* (Cate, 1978), length 13.5 mm, 200 m depth off Key Largo, Florida Keys; **I, J**= *Cyphoma alleneae* Cate, 1973, length 33.5 mm, Missouri Key, Florida Keys; **K**= *Arene vanhyningi* Rehder, 1943, holotype, diameter 11 mm, Sand Key, south of Key West, Florida Keys (differs from the widespread *A. cruentata* by being a much more flattened shell with smooth and unsculptured spire whorls, and in having a dark khaki-gray color on the spire whorls). For an overview of the Keys ovulid fauna, and illustrations of the closely-related *Cyphoma gibbosum*, *C. mcgintyi*, and *C. signatum*, see Petuch and Myers, 2014: 97-99.

Figure 3.5 Floridian Subprovince Endemic Species.
A= *Leucozonia jacarusoi* Petuch, 1987, holotype, length 24 mm, at 200 m depth on coralline algae, Dry Tortugas, Florida Keys (differs from the shallow water *L. nassa* by having a much longer siphonal canal, reduced body whorl sculpture, a much weaker peripheral cord and labial tooth, and in being white or yellow in color); **B=** *Vokesimurex marcoensis* (Sowerby III, 1900), length 27.9 mm, orange color form, Middle Torch Key, Florida Keys; **C, D=** *Vokesimurex marcoensis* (Sowerby III, 1900), length 28.4 mm, in Turtle Grass, Middle Torch Key, Florida Keys (compare with *Vokesimurex rubidus*, Figure 1.4, Chapter 1); **E, F=** *Falsuszafrona taylorae* (Petuch, 1987), length 11 mm, in Turtle Grass, Joe Kemp Key, Florida Bay; **G, H=** *Uzita swearingeni* Petuch and Myers, 2014, length 8.5 mm, Middle Torch Key, Florida Keys; **I, J=** *Uzita websteri* Petuch and Myers, 2014, length 8 mm, Howe Key, Florida Keys.

Figure 3.6 Floridian Subprovince Endemic Species.
A, B= *Pusula juyingae* Petuch and Sargent, 2011, length 15 mm, Elliot Key, Florida Keys (original description reprinted in the Systematic Appendix at the end of this book; incorrectly referred to as "*Pusula costispunctata*", which is actually an Eastern Pacific species); **C, D=** *Trochomodulus calusa* (Petuch, 1988), width 13 mm, Middle Torch Key, Florida Keys; **E=** *Trochomodulus calusa* (Petuch, 1988), width 12 mm, with attached commensal Slipper Shell (*Crepidula ustulatulina*), Joe Kemp Key, Florida Bay; **F, G=** *Lithopoma americana* (Gmelin, 1791), height 24 mm, Middle Torch Key, Florida Keys; **H, I=** *Haminoea taylorae* Petuch, 1987, length 9 mm, Joe Kemp Key, Florida Bay; **J=** *Haminoea taylorae* Petuch, 1987, length 10 mm, Joe Kemp Key, Florida Bay.

Figure 3.7 Floridian Subprovince Endemic Species.
A, B= *Cancellaria adelae* Pilsbry, 1940, length 39 mm, Middle Torch Key, Florida Keys; **C, D=** *Amathina pacei* (Petuch, 1987), width 9.5 mm, found attached to a *Caribachlamys* scallop on Pickles Reef, Plantation Key, Florida Keys; **E, F=** *Dibaphimitra florida* (Gould, 1856), length 55 mm, in an octopus midden on Pickles Reef, Plantation Key, Florida Keys; **G, H=** *Bulla frankovichi* Petuch and Sargent, 2011, length 10.2 mm, in Turtle Grass, Middle Torch Key, Florida Keys; **I=** *Bulla frankovichi* Petuch and Sargent, 2011, length 9 mm, Middle Torch Key, Florida Keys.

Figure 3.8 Floridian Subprovince Endemic Muricids.
A, B= *Dermomurex pacei* Petuch, 1988, length 17 mm, found within cavities in a gorgonian holdfast, off Pickles Reef, Plantation Key, Florida Keys; **C, D=** *Coralliophila kalafuti* (Petuch, 1987), length 12 mm, on *Porites furcata* coral on Pickles Reef, Plantation Key, Florida Keys; **E, F=** *Favartia pacei* Petuch, 1988, length 15.6 mm, in Yellow Mussel (*Brachidontes modiolus*) bed, Missouri Key, Florida Keys; **G, H=** *Murexiella caitlinae* Petuch and Myers, 2014, holotype, length 30 mm, in Yellow Mussel (*Brachidontes modiolus*) bed, Missouri Key, Florida Keys.

Figure 3.9 Endemic Muricids from the deep reefs of the Floridian Subprovince.
A, B= *Murexiella kalafuti* Petuch, 1987, length 15 mm, 50 m depth off the Dry Tortugas, Florida Keys; **C, D=** *Dermomurex (Trialatella) glicksteini* Petuch, 1987, length 12 mm, trawled from 100 m depth off Big Pine Key, Florida Keys; **E=** *Dermomurex (Trialatella) glicksteini* Petuch, 1987, holotype, length 13 mm, off Palm Beach, Florida; **F=** *Timbellus guesti* (Harasewych & Jensen, 1979), paratype USNM 783318, 20 mm, 150 fathoms off Key West, Florida; **G, H=** *Favartia goldbergi* Petuch and Sargent, 2011, length 10.5 mm, on carbonate rubble, 100 m depth off Palm Beach Island, Florida (original description reprinted in the Systematic Appendix at the end of this book); **I, J=** *Pygmaepterys richardbinghami* (Petuch, 1987), holotype, length 20 mm, off Palm Beach Island, Florida.

Figure 3.10 *Cinctura tortugana* from the deep water off the Florida Keys.
A, B= *Cinctura tortugana* (Hollister, 1957), length 70.8 mm; **C=** *Cinctura tortugana* (Hollister, 1957) length 97.8 mm; **D=** *Cinctura tortugana* (Hollister, 1957) length 96.5 mm.. Captured in deep water lobster traps, set in 100 m depth off Garden Key, Dry Tortugas, Florida Keys.

Figure 3.11 *Fulguropsis keysensis* from the Florida Keys Turtle Grass Beds.
A, B= *Fulguropsis keysensis* Petuch, 2013, length 87 mm, 2 m depth off Howe Key, Lower Florida Keys; **C, D=** *Fulguropsis keysensis* Petuch, 2013, length 68, 2 m depth in sand near sponge and *Porites* coral bioherms, Middle Torch Key, Lower Florida Keys; **E, F=** *Fulguropsis keysensis* Petuch, 2013, length 65 mm, 2 m depth off No Name Key, Lower Florida Keys. This heavily-ribbed species occurs together with the smooth-sculptured *Fulguropsis pyruloides* only along a small stretch of coast in the Ten Thousand Islands.

Figure 3.12 *Melongena (Rexmela) bicolor* from intertidal areas of the Florida Keys.
A, B= *Melongena (Rexmela) bicolor* (Say, 1827), length 36 mm, Missouri Key, Florida Keys; **C, D**= *Melongena (Rexmela) bicolor* (Say, 1827), length 35 mm, West Summerland Key, Florida Keys; **E, F**= *Melongena (Rexmela) bicolor* (Say, 1827), length 36 mm, Big Pine Key, Florida Keys. All three specimens were collected in tide pools on exposed rocky terraces adjacent to the island shorelines.

Figure 3.13 ***Melongena (Rexmela) bicolor*** **forms** ***minor*** **and** ***estephomenos*****.**
A, B= *Melongena (Rexmela) bicolor* form *minor* (Sowerby III, 1879), length 23 mm, Middle Torch Key, Florida Keys; **C=** *Melongena (Rexmela) bicolor* form *minor* (Sowerby III, 1879), length 22 mm, Little Torch Key, Florida Keys; **D, E=** *Melongena (Rexmela) bicolor* form *estephomenos* (Melvill, 1881), length 33 mm, Long Key, Florida Keys; **F=** *Melongena (Rexmela) bicolor* form *estephomenos* (Melvill, 1881), length 34 mm, Long Key, Florida Keys.

Figure 3.14 *Scaphella elizabethae* from the Dry Tortugas.
A, B= *Scaphella elizabethae* Petuch and Sargent, 2011, length 110 mm, 10 m depth off Garden Island, Dry Tortugas, Florida Keys; **C, D=** *Scaphella elizabethae* Petuch and Sargent, 2011, holotype, length 106.8 mm (USNM 1152529, National Museum of Natural History, Smithsonian Institution; original description reprinted in the Systematic Appendix at the end of this book), length 106.8 mm, 7 m depth off the Dry Tortugas, Florida Keys. Endemic to the Dry Tortugas and the extreme southwestern edge of the Florida Platform.

Figure 3.15 ***Scaphella glicksteinorum*** **from the Florida Coral Reef Tract.**
A, B= *Scaphella glicksteinorum* Petuch and Berschauer, 2023, holotype, length 64 mm, 5 m depth off Big Pine Key, Florida Keys; **C, D=** *Scaphella glicksteinorum* Petuch and Berschauer, 2023, length 63 mm, 3 m depth off Big Pine Key, Florida Keys. This dwarf volute is endemic to shallow water areas along the Florida Coral Reef Tract, from Palm Beach to Key West, where it prefers pockets of clean carbonate sand among living corals.

Figure 3.16 *Americoliva bollingi* from the Floridian Subprovince.
A, B= *Americoliva bollingi* (Clench, 1943), dark color form, length 49 mm, off Pickles Reef, Florida Keys; **C, D=** *Americoliva bollingi* (Clench, 1943), brown-banded color form, length 45 mm, off Soldier Key, Biscayne Bay; **E, F=** *Americoliva bollingi* (Clench, 1943), light color form, length 46 mm, off Big Pine Key, Florida Keys. Previously considered to be a subspecies of *A. nivosa*, which is a Bermuda endemic species, but here treated as a full valid species.

Figure 3.17 *Gradiconus burryae* from the Florida Keys Turtle Grass Beds.
A, B= *Gradiconus burryae* (Clench, 1942), length 43 mm, dark color form; **C, D**= *Gradiconus burryae* (Clench, 1942), length 37.8 mm, light color form with mottled pattern; **E, F**= *Gradiconus burryae* (Clench, 1942), length 35.5 mm, dark color form. All three specimens were collected in sand patches adjacent to Turtle Grass beds, in 2 m depth off Missouri Key, Florida Keys.

Figure 3.18 *Gradiconus mazzolii* from the Florida Keys Sponge Beds.
A, B= *Gradiconus mazzolii* Petuch and Sargent, 2011, length 22 mm; **C, D=** *Gradiconus mazzolii* Petuch and Sargent, 2011, length 21 mm; **E, F=** *Gradiconus mazzolii* Petuch and Sargent, 2011, length 20.8 mm. All three specimens were collected on sponge and Finger Coral (*Porites furcata*) bioherms in 1-2 m depth off Middle Torch Key, Lower Florida Keys. Although considered to be a synonym of *Gradiconus burryae* by some workers, the consistent differences in shell shape between the two species, along with the different animal colors and radular teeth shapes, show that *burryae* and *mazzolii* are two distinct species.

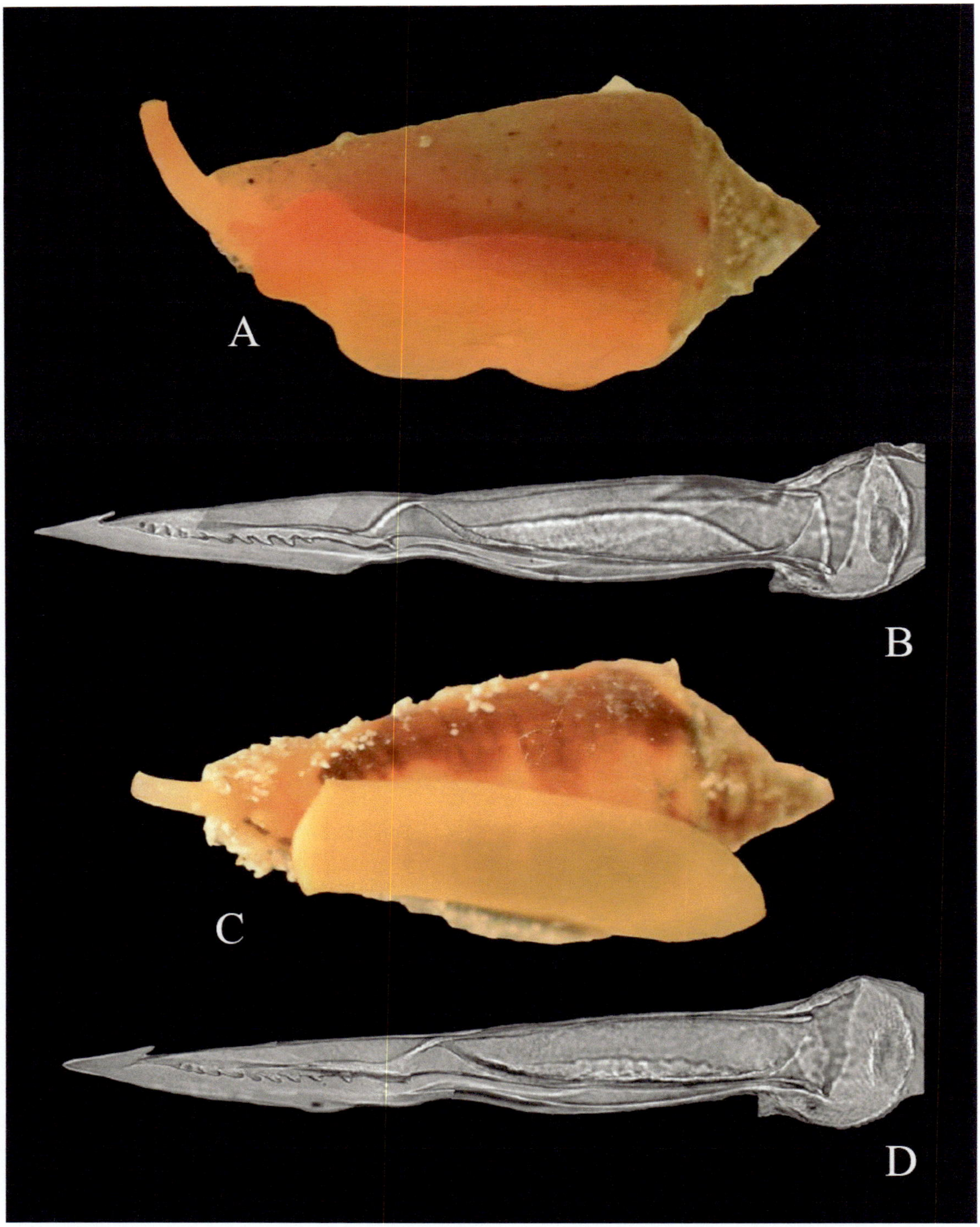

Figure 3.19 Comparison of the living animals and radular teeth of *Gradiconus burryae* and *Gradiconus mazzolii*. **A**= *Gradiconus burryae* (Clench, 1942), showing the bright orange-red color of the living animal; **B**= radular tooth of *Gradiconus burryae* (Clench, 1942); **C**= *Gradiconus mazzolii* (Petuch and Sargent, 2011), showing the bright yellow color of the living animal; **D**= radular tooth of *Gradiconus mazzolii* (Petuch and Sargent, 2011). As can be seen here, the radular tooth of *G. mazzolii* is narrower and more slender than that of *G. burryae*, has a much smaller and less-developed lateral blade, and a proportionally larger terminal hook. The animal of *G. burryae* is orange-red while the animal of *G. mazzolii* is a bright lemon-yellow. Radular photomicrographs courtesy of Dr. Manuel Tenorio; photos of the living animals courtesy of William Bennight.

Figure 3.20 *Gradiconus patglicksteinae* from the Florida Reef Tract.
A= *Gradiconus patglicksteinae* (Petuch, 1987), holotype, length 25 mm, off Palm Beach, Florida (protoconch and early whorls eroded off); **B**= *Gradiconus patglicksteinae* (Petuch, 1987), length 27 mm, Dry Tortugas, Florida Keys; **C, D**= *Gradiconus patglicksteinae* white color form named *tranthami* Petuch, 1995, length 26 mm, Pickles Reef, Florida Keys; **E, F**= *Gradiconus patglicksteinae* pink and orange color form named *antoni* Cargile, 2011, length 31 mm, Dry Tortugas, Florida Keys. This highly-variable cone lives in pockets of clean sand near living corals and is found along the entire Florida Coral Reef Tract, from Palm Beach to the Dry Tortugas. Besides color, the most variable character seen in this species is the height of the spire, which varies from flattened (like the holotype) to elevated (like the *tranthami* form illustrated here).

Figure 3.21 *Jaspidiconus* species of the Floridian Subprovince.
A, B= *Jaspidiconus vanhyningi* (Rehder, 1944), length 15 mm, off Pickles Reef, Florida Keys; **C=** *Jaspidiconus pealii* (Green, 1830) length 13.2 mm, typical color form, in Turtle Grass, Middle Torch Key, Florida Keys; **D=** *Jaspidiconus pealii* (Green, 1830) length 16 mm, orange color form, in Turtle Grass, Middle Torch Key, Florida Keys; **E, F=** *Jaspidiconus fluviamaris* Petuch and Sargent, 2011, holotype, length 16 mm, Hillsboro Inlet, Florida (also found on coral reefs in the Florida Keys and Dry Tortugas); **G, H=** *Jaspidiconus pealii* (Green, 1830), length 13 mm, banded color form, in Turtle Grass, Middle Torch Key, Florida Keys; **I, J=** *Jaspidiconus acutimarginatus* (Sowerby II, 1866), length 19.7 mm, Ohio Key, Florida Keys (see Eason, 2018).

Figure 3.22 ***Tuckericonus caribaeus*** **from the deep reefs of the Floridian Subprovince.**
A, B= *Tuckericonus caribaeus* (Clench, 1942), golden-brown color form, length 19 mm, 20 m depth off Big Pine Key, Florida Keys; **C, D=** *Tuckericonus caribaeus* (Clench, 1942), length 22 mm, 50 m depth off Palm Beach Island, Florida (white color form similar to the holotype of *T. caribaeus*); **E=** *Tuckericonus caribaeus* (Clench, 1942), brown zig-zag pattern, length 15 mm, 20 m depth off Soldier Key, Biscayne Bay, Florida; **F=** *Tuckericonus caribaeus* (Clench, 1942), golden form with white zig-zags, length 15.5 mm, 20 m depth off Key West, Florida.

Figure 3.23 *Tuckericonus flamingo* from the deep reefs of the Floridian Subprovince.
A, B= *Tuckericonus flamingo* (Petuch, 1980), length 21 mm, 100 m depth off Palm Beach Island, Florida; **C, D**= *Tuckericonus flamingo* (Petuch, 1980), length 19 mm, 100 m depth off Palm Beach Island, Florida; **E, F**= *Tuckericonus flamingo* (Petuch, 1980), length 21 mm, 200 m depth off the Dry Tortugas, Florida; **G**= *Tuckericonus flamingo* (Petuch, 1980), color form with prominent white zig-zag flammules, length 21 mm, 150 m depth off the Dry Tortugas, Florida.

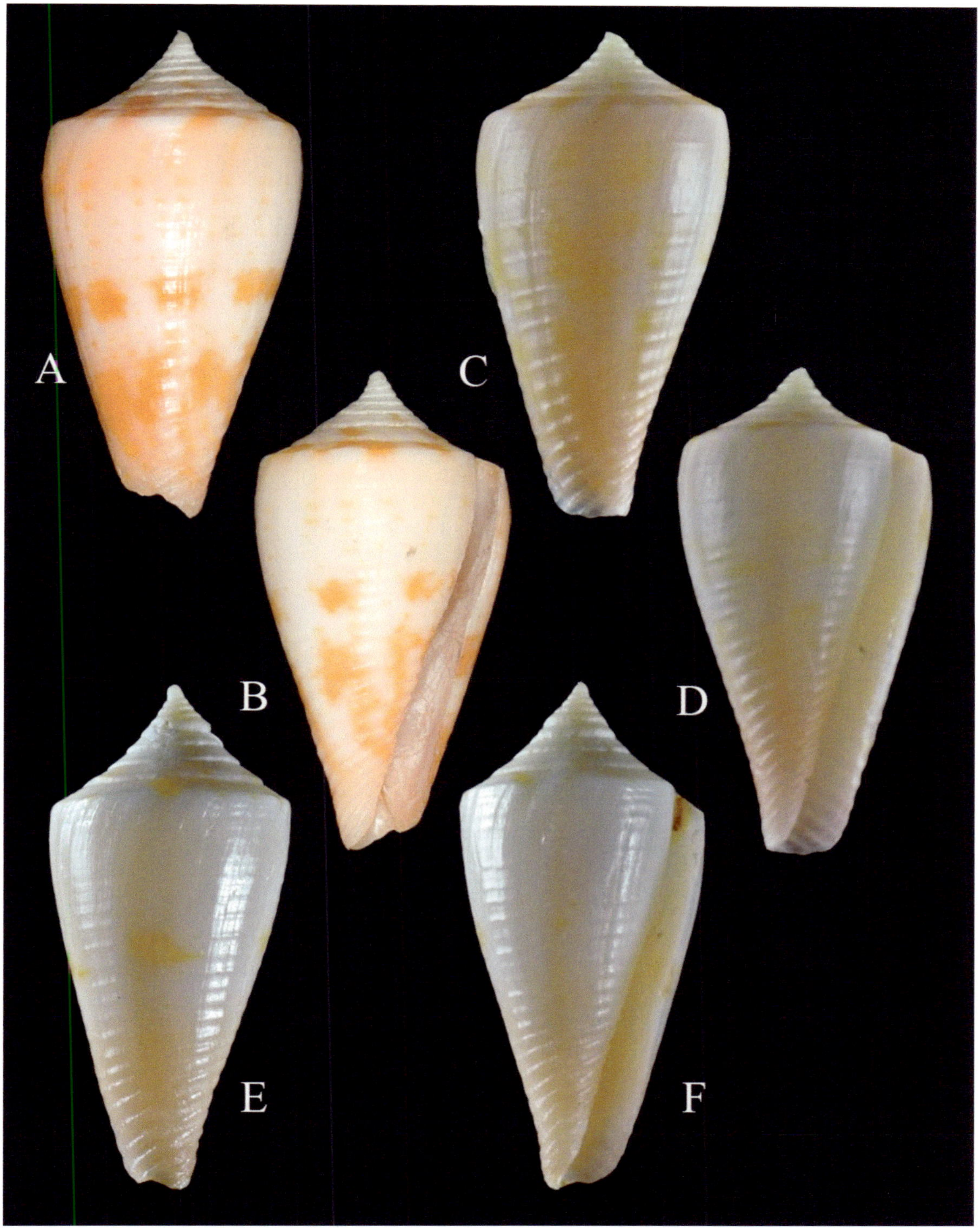

Figure 3.24 *Kellyconus binghamae* from the deep reefs of the Floridian Subprovince.
A, B= *Kellyconus binghamae* (Petuch, 1987), holotype, length 18.0 mm, off Palm Beach, Florida; **C, D**= *Kellyconus binghamae* (Petuch, 1987), length 14.0 mm, 100 m depth off the Dry Tortugas, Florida Keys; **E, F**= *Kellyconus binghamae* (Petuch, 1987), length 13.0 mm, 100 m depth west of Naples, Florida.

Figure 3.25 *Dauciconus glicksteini* from the deep reefs of the Floridian Subprovince.
A, B= *Dauciconus glicksteini* (Petuch, 1987), length 23.0 mm, 80 m depth off Big Pine Key, Florida Keys; **C, D**= *Dauciconus glicksteini* (Petuch, 1987), length 20.0 mm, 100 m depth off the Dry Tortugas, Florida Keys; **E, F**= *Dauciconus glicksteini* (Petuch, 1987), length 16.9 mm, 100 m depth off Palm Beach, Florida.

Figure 3.26 Deep Neritic Endemic Mollusks from the Floridian Subprovince.
A, B= *Prunum frumari* Petuch and Sargent, 2011, length 11.0 mm, 100 m depth off Key West, Florida; **C, D=** *Bartschia frumari* Garcia, 2008, holotype, length 73.8 mm, 100 m depth off Key West, Florida; **E, F, G=** *Arca rachelcarsonae* Petuch and Myers, 2014, length 23 mm, 80 m depth off Key Largo, Florida Keys; **H, I=** *Caribachlamys mildredae* (Bayer, 1941), length 43 mm, 50 m depth off Big Pine Key, Florida; 15.8 mm, 120 m; **J=** *Tritonoharpa janowskyi* Petuch and Sargent, 2011, length 15.8 mm, 100 m depth off Palm Beach, Florida.

Figure 3.27 *Fasciolaria bittneri* from deep reefs of the Floridian Subprovince.
A, B= *Fasciolaria bittneri* Petuch and Berschauer, 2023, holotype, length 115.1 mm; **C, D=** *Fasciolaria bittneri* Petuch and Berschauer, 2023, red color form, length 115.2 mm; **E, F=** *Fasciolaria bittneri* Petuch and Berschauer, 2023, length 117.9 mm. All specimens were collected from crab traps set at 50 m depth off the Dry Tortugas, Florida Keys. Also collected off Palm Beach County. The early whorls are always bright orange and bear prominent strong knobs.

Figure 3.28 Olive Shells from the Bathyal Zone of the Floridian Subprovince.
A, B= *Americoliva matchetti* Petuch and Myers, 2014, length 18 mm, 100 m depth off Big Pine Key, Florida Keys; **C, D=** *Americoliva matchetti* Petuch and Myers, 2014, length 15 mm, 100 m depth off the Dry Tortugas, Florida Keys; **E, F=** *Americoliva recourti* Petuch and Myers, 2014, length 23 mm, 100 m depth off Key Largo, Florida Keys; **G, H=** *Americoliva recourti* Petuch and Myers, 2014, length 22 mm, 100 m depth off Key West, Florida Keys.

Figure 3.29 *Clenchina atlantis* from the Floridian Subprovince Bathyal Zone.
A, B= *Clenchina atlantis* (Clench, 1946), holotype, length 97.5 mm, trawled from 400 m depth between Matanzas, Cuba and Key West, Florida. This holotype specimen is missing from the Harvard Museum of Comparative Zoology collection and has apparently been stolen (a "kleptotype"). The smaller specimen shown here could be designated as a neotype if the holotype is never recovered; **C, D=** *Clenchina atlantis* (Clench, 1946), length 53 mm, trawled in 421 m depth in the Straits of Florida south of Key West, *R/V Atlantis* station 2999 (collected March 17, 1938; from the M.L. Jaume collection).

Figure 3.30 ***Clenchina florida*** **from the Floridian Subprovince Bathyal Zone.**
A, B= *Clenchina florida* (Clench and Aguayo, 1940), length 55 mm, 262 m depth off Key Largo, Florida Keys; **C, D=** *Clenchina florida* (Clench, 1940), length 57 mm, 300 m depth off Key West, Florida Keys; **E, F=** *Clenchina florida* (Clench 1940), gerontic specimen, length 71 mm, 250 m depth off Key West, Florida Keys. Restricted to the Florida Keys and Dry Tortugas, where it lives together with the similar and wide-ranging *Clenchina dohrni*.

Figure 3.31 *Clenchina rundoi* from the Floridian Subprovince Bathyal Zone.
A, B= *Clenchina rundoi* Petuch and Berschauer, 2023, length 40.3 mm; **C, D=** *Clenchina rundoi* Petuch and Berschauer, 2023, holotype, length 58 mm (broken); **E, F=** *Clenchina rundoi* Petuch and Berschauer, 2023, length 41.5 mm. All three specimens were dredged from 350-400 m depth off the Dry Tortugas, Florida Keys, Monroe County, Florida, by deep water shrimpers.

Figure 3.32 ***Rehderia georgiana*** **from the Outer Neritic and Bathyal Zones of the Georgian and Floridian Subprovinces. A, B=** *Rehderia georgiana* (Clench, 1946), length 80 mm, 150 m depth off Palm Beach Island, Florida; **C, D=** *Rehderia georgiana* (Clench, 1946), length 78 mm, 200 m depth off Key Largo, Florida Keys.

Figure 3.33 *Rehderia schmitti* from the Bathyal Zone of the Floridian Subprovince.
A, B= *Rehderia schmitti* (Bartsch, 1931), length 82 mm, 300 m depth northwest of the Dry Tortugas, Florida Keys; **C, D**= *Rehderia schmitti* (Bartsch, 1931), length 84 mm, 350 m depth off Key West, Florida Keys.

Figure 3.34 *Clenchina carrieae* from the Bathyal Zone of the Straits of Florida.
A, B= *Clenchina carrieae* Petuch and Berschauer, 2023, holotype, length 42 mm, dredged by the R/V *Combat* (station 444), from 550 m depth east of Key Largo, Florida; **C, D=** *Clenchina carrieae* Petuch and Berschauer, 2023, length 40 mm, trawled by deep water *Glyphograngon* "Royal Red" shrimpers, from 700 m depth in the Straits of Florida, south of Key West, Florida; **E, F=** *Clenchina carrieae* Petuch and Berschauer, 2023, length 44 mm, Trawled from 650 m depth off Key West, Florida.

Figure 3.35 Deep reef Caribbean Province species that also occur in the Gulf of Mexico.
A= *Cyphoma intermedia* (Sowerby I, 1828), length 26.5 mm; **B**= *Oniscidia dennisoni* (Reeve, 1842), length 36 mm; **C**= *Babelomurex pacei* (Petuch, 1987), length 31 mm; **D**= *Ranularia rehderi* (Verrill, 1950), apertural view, length 57 mm; **E**= *Jaspidiconus mindanus* (Hwass, 1792), length 35 mm; **F**= *Atlanticonus granulatus* (Linnaeus, 1758), length 65 mm (Floridian and Yucatanean Subprovinces only); **G**= *Phyllonotus oculatus* (Reeve, 1845), length 83 mm; **H**= *Fenimorea sunderlandi* (Petuch, 1987), holotype, length 39.8 mm, from 150 m depth, 50 km south of Apalachicola, Franklin County, Florida; **I**= *Proadusta surinamensis* (Perry, 1811), length 26.5 mm, found on deep reefs in the Floridian and Yucatanean Subprovinces (the giant globose form *aubryana* has been found off Cozumel Island; see Petuch, 1987, plate 6, figures 3 and 4, which illustrates a 45 mm specimen).

Figure 3.36 *Naria turdus*, an introduced Indian Ocean species that has recently been discovered in the Floridian and Yucatanean Subprovinces.

A, B= *Naria turdus* (Lamarck, 1810), length 34.7 mm, Malmok, Aruba, Netherlands Antilles; **C, D=** *Naria turdus* (Lamarck, 1810), length 34.5 mm, Malmok, Aruba, Netherlands Antilles. This common and widespread cowrie is naturally found throughout the tropical Indian Ocean, from East Africa to Thailand, and is also abundant in the Persian Gulf and the entire Red Sea. It was discovered, several decades ago, that this attractive cowrie had circumvented the Suez Canal ecological barrier and had become established in the eastern Mediterranean Sea (referred to as a "Lessepsian Migrant"). Mediterranean ecophenotypical variants were given the form name *micheloi* by Chiapponi in 2009. Within the last ten years, *Naria turdus* has been collected in the Caribbean Sea and was found on Aruba and Curaçao Islands, the coast of Venezuela, Puerto Rico, and the Dominican Republic. These specimens appear to have been introduced either as pelagic larvae or juvenile individuals that had survived within the bilge water of petroleum super-tankers that arrived from the Persian Gulf and Gulf of Aden regions. This non-indigenous cowrie has recently been collected in Palm Beach, Fort Lauderdale, and Key West in southern Florida and also near Cancun, Yucatan Peninsula, Mexico, showing that it is rapidly expanding its range within the western Atlantic. Since the Palm Beach and Key West areas represent the extreme northeastern and western ends of the Floridian Subprovince, and Cancun represents the extreme southwestern end of the Yucatanean Subprovince, this non-indigenous invader can now be considered to be a component of the Carolinian Province malacofauna. The western Atlantic ecophenotypical variant of *Naria turdus* was given the form name *gabrielli* by Ros in 2022 and this infraspecific taxon would also apply to the Gulf of Mexico specimens. For details on the discovery of *Naria turdus* in the western Atlantic, see Dekkers and Ros, 2022 and Oleinik *et al.*, 2023.

View of a Dwarf Mangrove forest, composed of naturally-occurring bonsai Red Mangroves (*Rhizophora mangle*), growing along the northern side of the Saddlebunch Keys, Lower Florida Keys. These stunted trees are growing on the dense oolitic limestone of the Pleistocene Miami Formation, in only a few centimeters of carbonate sediments, and cannot develop to their full potential size. The open areas between the mini-forests support a rich molluscan fauna composed of beds of the Gulf endemic bivalve *Polymesoda floridana*, the dwarf melongenid *Melongena (Rexmela) bicolor* form *minor*, potamidids such as *Cerithidea costata* and *Cerithideopsis scalariformis*, and air-breathing ellobiids.

CHAPTER 4. Endemic Mollusks of the Suwannean Subprovince

Extending from the Ten Thousand Islands of southwestern Florida northward to the Mississippi River Delta, the Suwannean Subprovince is the most ecologically-complex of the four Gulf subprovinces (Petuch and Berschauer, 2021a). Named for the Suwannee River of northwestern Florida, this eastern Gulf subprovince also spans the entire wide Florida Platform and is, spatially, the largest Carolinian subdivision in the Gulf of Mexico. The Suwannean biotic area contains the largest number of environmental types, encompassing such varied biotopes as emergent Mangrove forests, Cordgrass meadows, muddy estuaries, oyster banks, quartz sand beaches and sand flats, sponge-covered carbonate banks, and deep water coralline algal beds. Because of this great diversity of habitats, the Suwannean Subprovince has a combined species-rich molluscan fauna with a very high level of endemism. Many of the most conspicuous endemic macromollusks are shown in the iconography at the end of this chapter, and these can be used as index taxa for defining the biogeographical limits of the subprovince.

Since much of the area of the Suwannean Subprovince is under the influence of the Gulf Loop Current (see Chapter 2), the offshore water remains much warmer than the water adjacent to the Florida coastline The entire broad Florida Platform is continuously bathed in subtropical oceanic conditions, allowing for scattered growth of hermatypic corals and the accumulation of carbonate sediments. Closer inshore, the carbonate sediments give way to broad swaths of quartz sand and much of the coastline is dominated by white quartz sand beaches. During the winter months, these coastal areas are subject to very cold air conditions, often dropping to near-freezing temperatures for weeks at a time. The combination of the sterile quartz sand and cold winter water temperatures has resulted in the formation of an impoverished coastal malacofauna, with only a fraction of the biodiversity seen in the adjacent Floridian Subprovince (see Petuch and Berschauer, 2022 for a biotic survey of the Ten Thousand Islands of the southeastern Gulf of Mexico). Because of the great disparity in water temperatures along the coastal areas of the Suwannean Subprovince, two "evolutionary hot spots" have developed, with each containing a high percentage of endemism. These highly-localized pockets of speciation, referred to here as "infraprovinces" (Petuch, 2013; Petuch and Berschauer, 2021a), represent areas of special evolution and include the **Chokoloskean Infraprovince** in the south (Ten Thousand Islands) and the **Apalachicolan Infraprovince** in the north (Florida Panhandle). These two infraprovincial bioregions will be discussed in detail in the following sections of this chapter.

Besides housing a large compliment of Carolinian Province index taxa, including gastropods such as the muricid *Hexaplex fulvescens*, the cypraeid *Macrocypraea (Lorenzicypraea) cervus*, the cassid *Cassis spinella*, and the cone shell *Kohniconus delessertii*, and bivalves such as the pectinid *Nodipecten fragosus* and the venerid *Mercenaria campechiensis*, the shallow coastal areas of the Suwannean Subprovince also support a rich endemic malacofauna. Some of these classic widespread eastern Gulf

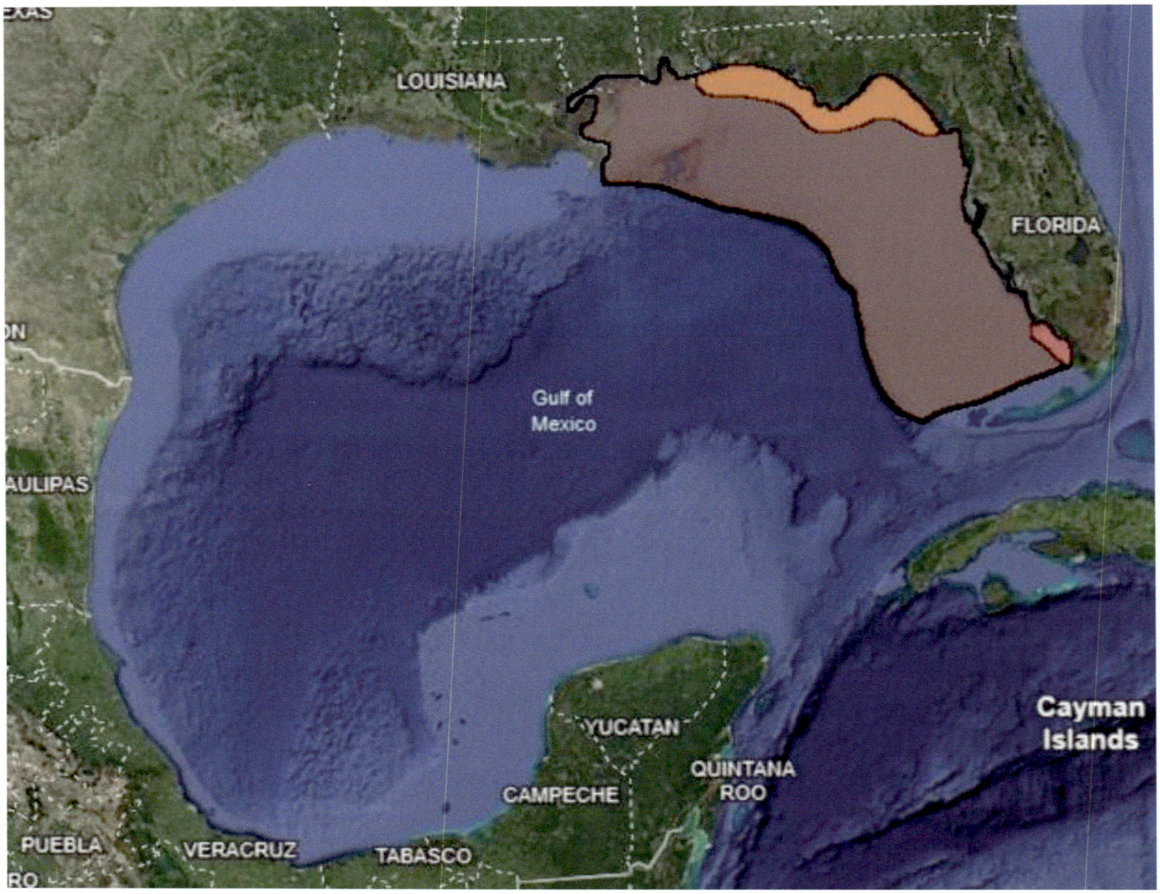

Figure 4.1 Map showing the geographical extent of the Suwannean Subprovince (brown, rose and orange shading) of the Carolinian Province, showing the wide Florida Platform and the distributions of the **Chokoloskean Infraprovince** (rose shading) and the **Apalachicolan Infraprovince** (orange shading).

endemics include gastropods such as the calliostomatid *Calliostoma tampaense*, the muricids *Vokesinotus perrugatus* and *Eupleura tampaense* (all three shown in Figure 4.6), the busyconid *Fulguropsis pyruloides* (Figure 4.10), the conids *Gradiconus floridanus* (Figures 4.12 and 4.13) and *Jaspidiconus stearnsi* (Figure 4.14), and the terebrid *Neoterebra vinosa* (Figure 4.6), and bivalves such as the sea grass pectinid *Argopecten irradians taylorae* (Figure 4.15). Of special interest along the intertidal areas of the Suwannean Subprovince is an array of forms and subspecies of the King's Crown Conch, *Melongena (Rexmela) corona* (Figures 4.7, 4.8, and 4.9), unlike anything else found in the Gulf. This large, abundant carnivore feeds primarily on oysters, but will feed on other mollusks (including the young of its own species) and carrion. Depending on the type of substrate (biotope), salinity of the sea water, and geographical area, *M. (Rexmela) corona* can assume a bewildering variety of morphological forms, including dwarf forms with reduced shoulder spines (the form named *belknapi*; Figure 4.8), forms with three rows of low knobs (the form named *trinodulosa*; Figure 4.8), dwarf forms with extreme spine development (the form named *perspinosa*; Figure 4.9), giant specimens with reduced spine development (the form named *mulletensis*), and specimens that are completely devoid of spines (the form named *inspinata*). A true subspecies of *M. (Rexmela) corona* has evolved in the northern section of the Suwannean Subprovince, due to partial geographic and oceanographic barriers that have developed along the northwestern Florida coast since the late Pleistocene. This large vase-shaped melongenid, *M. (Rexmela) corona johnstonei* (shown here 1n Figure 4.31), ranges from Cedar Key northward to Mobile Bay, Alabama and also has a number of morphological variants. One of these, the form named *incurvata* (Figure 4.32), is found primarily around Cedar Key and has shoulder spines that curl inwardly, almost touching the previous spire whorls. Another characteristic index taxon for the Suwannean Subprovince is *Americoliva sayana sarasotaensis* (Figure 4.11), an endemic subspecies of the United States East Coast *A. sayana sayana* (North Carolina south to Palm Beach, Florida; see Petuch and Berschauer, 2021a) that is only found along the west coast of Florida (Ten Thousand Islands to the Florida Panhandle). Since *A. sayana* is absent from the entire Floridian Subprovince, there is a wide biogeographical gap between the eastern Florida *A. sayana sayana* and the western Florida *A. sayana sarasotaensis*, with no genetic flow between the two subspecies. The carbonate and coral environments of the Florida Keys act as an ecological barrier.

Other gastropod groups have undergone incipient species radiations in the shallow water coastal areas of the Suwannean Subprovince. One of these is seen in the family Modulidae, where at least two species live in the Turtle Grass and Shoal Grass beds along the coast and in shallow bays and lagoons. Like *Melongena (Rexmela) corona*, these small algivores have diverged into a southern species, *Modulus floridanus* (Figure 4.6), which ranges from the Ten Thousand Islands northward to Cedar Key, and a similar-looking northern species, *Modulus danielsi* (Figure 4.30), which ranges from Cedar Key northward to the Mississippi River Delta. The newly-recognized *M. danielsi* also ranges westward to the coast of Texas, where rare specimens have been collected south of Galveston. Some groups of Suwannean gastropods have also evolved species complexes whose congeneric members are bathymetrically segregated; with a shallow water species, a mid-shelf neritic species, and a deep neritic or bathyal species. One of these bathymetric sibling complexes involves the pisaniid genus *Hesperisternia*: with both the widespread Carolinian Province *H. multangula* (southern Suwannean Subprovince) and the Suwannean endemic *H. grandana* (northern Suwannean Subprovince, Figure 4.30) living in shallow water; the Suwannean endemic *M. sulzyckii* living in offshore red algae beds (Figure 4.18); and the Suwannean endemic *H. harasewychi* (Figure 4.18) being restricted to deep neritic environments. Another cluster of bathymetrically-distributed sibling species is seen in the genus *Cyphoma* (family Ovulidae). These octocorallian-feeding gastropods have moved into two separate depth zones, with the wide-ranging Carolinian Province *Cyphoma mcgintyi* living on *Leptogorgia* Sea Whips in shallow water and the Suwannean endemic *Cyphoma lindae* (Figure 4.16) living on deep neritic antipatharian octocorals.

The Deep Water Seafloors and Coralline Algal Beds

Within the Suwannean Subprovince, the malacofauna with the highest level of biodiversity is found along the extreme western edge of the Florida Platform, in depths averaging between 100 and 200 m. At this depth range, the relatively flat sea floor is covered with a carbonate and quartz sand mix, often existing as a thin covering over a hard, dense limestone of Miocene age. This type of sea floor, which supports large biohermal assemblages of sponges and gorgonian octocorals, intergrades with broad swaths of sea floor that are covered exclusively with red coralline algae of the genera *Porolithon* and *Lithothamnion*. These coralline algal beds occur in the form of dense accumulations of small branched clumps called "rhodoliths" and typically form tightly interlocked beds that are meters in thickness. These rhodolithic beds create a porous substrate that is filled with a maze of interstices, offering shelter for a wide variety of highly-specialized endemic mollusks that are found nowhere else on Earth. Together, these two types of interfingering sea floors along the outer Neritic Zone support some of the highest biodiversity of marine invertebrates known from the Gulf of Mexico area.

The open sand and sponge bed sea floors of the outer Neritic Zone of the Suwannean Subprovince house a large number of endemic gastropods, many of which reach impressive sizes that are equivalent to those of their shallow water relatives. Some of the most conspicuous and important of these deep water gastropods include the thin and fragile, bright orange busyconid *Lindafulgur lyonsi* (Figure 4.24), the large pink and white, high-spired fasciolariid *Heilprinia lindae* (Figure 4.18), and the extremely slender, orange and brown-spotted terebrid, *Cinguloterebra lindae* (Figure 4.20). Characteristically, these large species occur together with a distinctive assemblage of smaller endemic gastropods which includes the tiny xenophorid carrier shell *Xenophora microdiscus* (Figure 4.16), the ornate turritellid *Torculoidella lindae* (Figure 4.18), the muricids *Phyllonotus whymani* (Figure 4.23) and *Vokesimurex lindajoyceae* (Figure 4.17), the volute *Clenchina marionae* (Figure 4.19), the small olivid *Americoliva sunderlandi* (Figure 4.20), the cancellariid *Cancellaria richardpetiti* (Figure 4.16), the marginellid *Prunum cassis*, the conids *Dauciconus aureonimbosus* (Figure 4.21) and *Gradiconus optabilis* (Figure 4.21), the drilliid *Fenimorea brunnescens* (Figure 4.20), and the turrid *Polystira sunderlandi* (Figure 4.20). These are all illustrated in the Iconography at the end of this chapter. The unusual nature of this molluscan assemblage is underscored by the co-existence of the busyconid *Lindafulgur lyonsi* with the slit shell *Perotrochus amabilis* and the abalone *Haliotis pourtalesii*. This is the only-known record of a busyconid whelk, an abalone, and a slit shell living together in the same ecosystem.

The *Porolithon* and *Lithothamnion* rhodolith beds along the edge of the continental shelf house an equally-unusual molluscan fauna that has a high level of endemism. Referred to as the "Tampan Secondary Relict Pocket" by the senior author (Petuch, 1987), this far-western region of the Florida Platform was recognized as containing numerous archaic genera. Primary among these relictual taxa is the modulid genus *Conomodulus*, where a single species, *C. lambi* (Figure 4.17), is known to inhabit the interstitial spaces within the rhodolith beds. This "living fossil" modulid, a remnant of older Miocene faunas, lives together with, and may be a prey item of, the five-spined muricid, *Chicoreus rachelcarsonae* (Figure 4.22). This delicate pink and orange murex shell is one of the real treasures of the coralline algal beds and closely resembles several archaic five-spined *Chicoreus* species known from the Pliocene and early Pleistocene fossil beds of southern Florida (see Petuch, 1994; Petuch and Berschauer, 2021b). Besides *C. rachelcarsonae*, no other *Chicoreus* species from Florida and the Caribbean has five spines per varix; all known living western Atlantic species have six frondose spines per varix and appear to represent a clade that is separate from the five-spined species complex represented by the living *C. rachelcarsonae* and the fossils *C. stephensae* and *C. susanae*. Another important inhabitant of the rhodolith beds is the spiny pectinid bivalve, *Lindapecten lindae* (Figure 4.25). This small ornate species, which is characterized by its striking pale salmon-pink base color and dark rose-red zig-zag markings and flattened valves, lives nestled within the interstices of the algal bed and is difficult to collect.

Endemism in the Chokoloskean Infraprovince

Of the shallow water intertidal coastal areas of the Suwannean Subprovince, the Ten Thousand Islands of extreme southwestern Florida (Collier and Monroe Counties) contain the most tropical environments found in the entire subprovince. Here, the solar-heated waters of the shallow southeasternmost Florida Platform keep the marine environment of this area warm throughout the year, allowing some tropical Floridian Subprovince endemics to have become established within the local ecosystems. Some of these Florida Keys migrants include the busyconid *Fulguropsis keysensis* (Figure 3.11), the columbellid *Falsuszafrona taylorae* (Figure 3.5), and the haminoeid *Haminoea taylorae* (Figure 3.6). The coastal area extending from Naples, Collier County, Florida southward to Cape Sable, Monroe County, Florida, and encompassing the entire Ten Thousand Islands archipelago, represents an "evolutionary hot spot", containing many important endemic mollusks and unique shallow water ecosystems. This geographically-small stretch of coastline, with its highly-localized malacofauna, was named the **Chokoloskean Infraprovince** (Figure 4.1) after the centrally-located Chokoloskee Island (see Petuch and Berschauer, 2022). The entire infraprovince is under the influence of the Everglades wetlands, with its immense fresh water effluent, and represents the area where the Everglades flows into the Gulf of Mexico. Because of this, the localities within the Chokoloskean Infraprovince have wildly-fluctuating salinities, depending on the seasonal rainfall input. Only the outermost islands, exposed directly to the open waters of the Gulf of Mexico, have stable year-round water chemistries. For a detailed overview of the Ten Thousand Islands, their unique environments, and their unusual molluscan faunas, see Petuch and Berschauer, 2022.

Figure 4.2 View of a large vermetoherm complex that is exposed at low tide on Turtle Key, Ten Thousand Islands. These "Worm Shell Reefs" are composed of massive aggregations of the vermetid gastropod, *Petaloconchus nigricans*, and grow along the seaward sides of the Ten Thousand Islands of southwestern Florida. Vermetoherms like this are unique to the Ten Thousand Islands and are characteristic of the Chokoloskean Infraprovince (see Petuch and Berschauer, 2022). A close-up of an aggregation of *Petaloconchus nigricans* is shown here in Figure 4.25.

Figure 4.3 View of three American White Ibis (*Eudocimus albus*) feeding on small mollusks and crustaceans at low tide on a vermetoherm off Turtle Key, Ten Thousand Islands, Florida.

The most distinctive and unusual ecosystem found within the Chokoloskean Infraprovince is one that is based entirely on massive "reef systems" composed entirely of aggregations of the suspension-feeding vermetid worm shell gastropod, *Petaloconchus nigricans* (Figures 4.2, 4.3, and 4.26). These extensive gastropod reefs, some covering many acres along the seaward sides of the outer islands, were named "**vermetoherms**" by Petuch and Berschauer (2022) and were found to have distinct growth forms that mimic their coral reef analogs. The largest and best-developed vermetoherms exhibit many similar geomorphological features, including an actively-growing fore reef area with flattened sheet-like aggregations (**stromatoverms**) and a quiet, back reef lagoon area with microatoll-like structures (**circoverms**). The nutrient-laden fresh water effluent from the Everglades mixes with the open oceanic sea water along the outer islands, creating the perfect conditions for the heavy plankton growth that supports the *Petaloconchus* vermetoherms. The stromatoverm and circoverm sections of the main vermetoherms also support a number of endemic Chokoloskean gastropods, including the pisaniid *Gemophos tinctus pacei* (which feeds on the living worm shells; Figure 4.27) and *Vermicularia fargoi owensi* (which lives in sponges on the vermetoherm platform; Figure 4.27). These occur with other classic Suwannean endemics such as the muricids *Eupleura tampaense* and *Vokesinotus perrugatus* and the calliostomatid *Calliostoma tampaense*, and form one of the strangest molluscan assemblages seen in the Gulf of Mexico and the only one known to have suspension-feeding aggregate gastropods as the foundation of the biotope.

The channels between the Ten Thousand Islands and the deeper offshore areas adjacent to the archipelago also house a rich malacofauna that contains many characteristic endemic mollusks. Primary among these is the small, endemic venerid clam, *Mercenaria browni* (Figure 4.29), which lives buried on intertidal sand flats, where it undergoes heavy predation from the predatory gastropods such as the large busyconid, *Sinistrofulgur sinistrum*, and the endemic Chokoloskean Infraprovince moon snail *Naticarius verae* (Figure 4.27). Of special interest in the offshore areas is the endemic fasciolariid, *Cinctura tortugana foxi* (Figure 4.28), a western Florida subspecies of the deeper water *C. tortugana* from the Floridian Subprovince (Florida Keys and Dry Tortugas). This Chokoloskean and western Florida subspecies is much more slender and has a longer siphonal canal than does the deep water nominate subspecies and also has a more vivid color pattern that is composed of either a deep rose-pink with reddish-brown flammules or a solid orange or red. All of these classic shallow water Chokoloskean endemic mollusks occur together with other typical Suwannean index species such as *Scaphella junonia*, *Fulguropsis pyruloides*, *Americolva sayana sarasotaensis*, *Jaspidiconus stearnsi*, and *Neoterebra vinosa.*

Endemism in the Apalachicolan Infraprovince

At the northern end of the Suwannean Subprovince, from roughly Cedar Key, Florida to Mobile Bay, Alabama, a second "evolutionary hot spot" has developed since the late Pleistocene. Centered on the Panhandle region of northern Florida, this ecologically-isolated area was named the **Apalachicolan Infraprovince** (Figure 4.1) for the city of Apalachicola, Franklin County, which is located in the geographic center of the infraprovincial bioregion (Petuch and Berschauer, 2021a).

Figure 4.4 View of a coastal lagoon area near Cedar Key, Levy County, Florida at low tide. Here large clumps of the Virginia Oyster, *Crassostrea virginica*, can be seen growing along the muddy shoreline adjacent to salt marshes dominated by Cordgrass (*Spartina alterniflora*) and small tree islands composed of pine trees and Cabbage Palmettoes. Areas like this are typical of the coastlines of northwestern Florida and the northern Gulf of Mexico. The Suwannean Subprovince endemic muricids, *Eupleura tampaensis* and *Vokesinotus perrugatus*, are primary predators on the oysters and often occur in large numbers.

Figure 4.5 Close-up view of a large clump of the Virginia Oyster, *Crassostrea virginica*, and a single individual of Johnstone's Crown Conch, *Melongena (Rexmela) corona johnstonei*, exposed at low tide on Cedar Key, Levy County, Florida.

Being located in the northeastern corner of the Gulf of Mexico, the Apalachicolan Infraprovince area is exposed to extremes in both air and water temperatures; with extremely hot summer temperatures, approaching the high tropics, and with very cold winters, with long periods of near-freezing temperatures and occasional heavy freezes during the peak winter months (discussed in Chapter 2). These non-tropical winter conditions preclude the development of any tropical or subtropical communities in this area and only the most physiologically-hardiest organisms can survive and flourish in a marine environment with such wide swings in temperature. For this reason, and also because of the fresh water effluent of the many river systems in the area (Figure 4.4), the Apalachicolan malacofauna has an impoverished appearance, lacking many of the common shallow water species seen farther south along southwestern Florida. Hardy Apalachicolan endemic taxa, such as the large melongenid *Melongena (Rexmela) corona johnstonei* (discussed earlier), along with the widespread Gulf muricid, *Stramonita canaliculata*, flourish in the exposed oyster beds in the myriad of lagoons and estuaries that typify the coastline of the Apalachicolan Infraprovince (Figure 4.5).

Despite this harsh temperature regime, a number of distinctive and characteristic endemic species have evolved within this area and these can be used as index species for demarcating the boundaries of the infraprovince. The *Halodule* and *Thalassia* sea grass beds of this area support several geographically-restricted endemic taxa such as the fasciolariid *Cinctura hunteria apalachee* (Figure 4.30), an endemic subspecies of the wide-ranging Carolinian *C. hunteria hunteria*, and the pisaniid *Hesperisternia grandana* (Figure 4.30). The latter species is particularly abundant in the **Big Bend Sea Grass Preserve** off Steinhatchee, Florida where it lives together with very large individuals of the Suwannean and Floridian Subprovince endemic scallop, *Argopecten irradians taylorae*. Sandy areas within shallow bays and coastal lagoons of the Florida Panhandle also house the olivid *Americoliva bollingi choctaw* (Figure 4.31), an isolated dwarf subspecies that is morphologically-closest to the larger *A. bollingi* from the Floridian Subprovince farther south. Probably the most beautiful and interesting of the Apalachicolan endemics is the large scaphelline volute, *Scaphella johnstoneae* (Figure 4.33), which lives in shallow offshore areas throughout the infraprovince, from the Mississippi River Delta east to the Big Bend area off Steinhatchee, Florida. This elongated and darkly-colored volute is frequently collected as by-catch in the nets of local shrimp and scallop fishermen, demonstrating that it is common in the subtidal and shallow neritic ecosystems.

Farther offshore from the Big Bend Sea Grass Preserve, along the northeastern portion of the broad West Florida Shelf, are wide open areas of clean sand sea floors. This relatively unstudied area has recently been found to house one of the largest-known species of Floridian *Jaspidiconus*, the heavily-sculptured *Jaspidiconus forticostatus* (Figure 5.34) (see Petuch and Berschauer, 2024 for details on the ecology of these offshore sand areas). This large ornate cone shell is most similar to the endemic Floridian Subprovince *Jaspidiconus acutimarginatus* and appears to represent a northeastern Gulf evolutionary offshoot of the Florida Reef Tract species. Besides containing broad expanses of clean sand, the offshore areas of the West Florida Shelf are also dotted with clumps of gorgonians and sponges, forming isolated bioherms. These are attached to the limestone platform (Eocene-Miocene age) that lies directly below the thin veneer of sand that typifies the area and support rarely-seen gastropods such as the small widespread abalone, *Haliotis pourtalesii*.

TABLE 5. Important Mollusks Endemic to the Suwannean Subprovince.

Gastropoda

Calliostomatidae
Calliostoma tampaense (Conrad, 1846)

Potamididae
Cerithidea turrita Stearns, 1872

Modulidae
Conomodulus lambi Petuch and Berschauer, 2023
Modulus danielsi Petuch and Berschauer, 2023
Modulus floridanus (Conrad, 1869)

Turritellidae
Torculoidella lindae Petuch, 1987
Vermicularia fargoi owensi Petuch and Myers, 2014

Ovulidae
Cyphoma lindae Petuch, 1987

Xenophoridae
Xenophora microdiscus Petuch, 1994

Naticidae
Naticarius verae (Rehder, 1947)

Muricidae
Chicoreus rachelcarsonae Petuch, 1987
Eupleura tampaense (Conrad, 1846)
Favartia lindae Petuch, 1987
Murexiella taylorae Petuch, 1987
Phyllonotus whymani Petuch and Sargent, 2011
Vokesimurex lindajoyceae Petuch, 1987
Vokesinotus perrugatus (Conrad, 1846)

Fasciolariidae
Aristofusus stegeri (Lyons, 1978)
Cinctura hunteria apalachee Petuch and Berschauer, 2020
Cinctura tortugana foxi Petuch and Berschauer, 2022
Heilprinia lindae Petuch, 1987

Melongenidae
Melongena (Rexmela) corona (Gmelin, 1791)
Melongena (Rexmela) corona form *belknapi* Petit de la Saussaye, 1850
Melongena (Rexmela) corona form *perspinosa* Pilsbry and Vanatta, 1934
Melongena (Rexmela) corona form *trinodulosa* Emery and Lermond, 1936
Melongena (Rexmela) corona johnstonei Clench and Turner, 1956
Melongena (Rexmela) corona johnstonei form *incurvata* Lermond, 1936

Busyconidae
Fulguropsis pyruloides (Say, 1822)
Lindafulgur lyonsi Petuch, 1987

Pisaniidae
Gemophos tinctus pacei Petuch and Sargent, 2011
Hesperisternia harasewychi Petuch, 1987
Hesperisternia sulzyckii Petuch and Myers, 2014
Hesperisternia grandana (Abbott, 1986)
Volutidae
Clenchina marionae Pilsbry and Olsson, 1953
Scaphella johnstoneae Clench, 1953
Olividae
Americoliva bollingi choctaw Petuch and Myers, 2014
Americoliva sayana sarasotaensis (Petuch and Sargent, 1986)
Americoliva sunderlandi (Petuch, 1987)
Marginellidae
Prunum cassis (Dall, 1889)
Cancellariidae
Cancellaria richardpetiti Petuch, 1987
Conidae
Dauciconus aureonimbosus Petuch, 1987
Gradiconus floridanus (Gabb, 1868)
Gradiconus floridanus form *floridensis* (Sowerby II, 1870)
Gradiconus optabilis (A. Adams, 1843)
Jaspidiconus forticostatus Petuch and Berschauer, 2024
Jaspidiconus stearnsi (Conrad, 1869)
Terebridae
Cinguloterebra lindae (Petuch, 1987)
Neoterebra vinosa (Dall, 1889)
Drilliidae
Fenimorea brunnescens Rehder, 1943
Turridae
Polystira sunderlandi Petuch, 1987
Bivalvia
Pectinidae
Argopecten irradians taylorae Petuch, 1987
Lindapecten lindae Petuch, 1995
Veneridae
Mercenaria browni Petuch and Berschauer, 2019

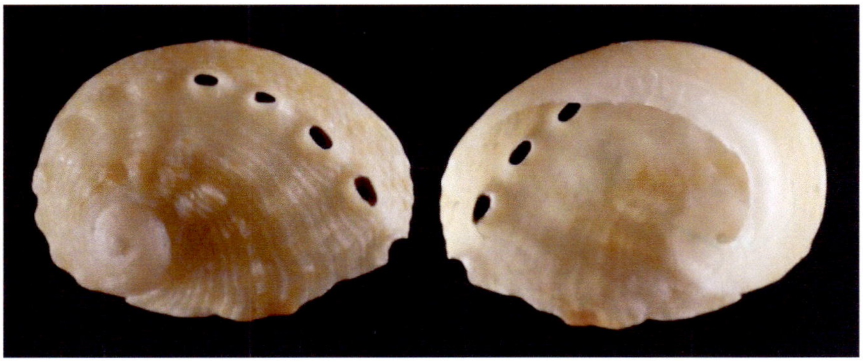

Haliotis pourtalesii Dall, 1881, width 7 mm; a rare and seldom-seen deep water abalone that is widespread across the Carolinian and Caribbean Molluscan Provinces, and is found throughout the Gulf of Mexico.

ICONOGRAPHY OF SUWANNEAN SUBPROVINCE ENDEMIC SPECIES

Figure 4.6 Widespread Suwannean Subprovince Endemic Taxa.
A= *Calliostoma tampaense* (Conrad, 1846), variant with a stepped spire, height 21 mm, Turtle Key, Ten Thousand Islands, Collier County, Florida; **B, C=** *Modulus floridanus* (Conrad, 1869), width 11 mm, Camp Lulu Key, Ten Thousand Islands, Collier County, Florida; **D=** *Cerithidea turrita* Stearns, 1872, length 15 mm, Steinhatchee, Taylor County, Florida. Found in tidal channels on intertidal mud flats and salt grass beds; **E=** *Neoterebra vinosa* (Dall, 1889), length 21 mm, Camp Lulu Key, Ten Thousand Islands; **F, G=** *Eupleura tampaense* (Conrad, 1846), length 27 mm, Goodland Bay, Collier County, Florida; **H, I=** *Vokesinotus perrugatus* (Conrad, 1846), length 23 mm, Camp Lulu Key, Ten Thousand Islands, Florida.

Figure 4.7 ***Melongena (Rexmela) corona*** **from southwestern Florida.**
A, B= *Melongena (Rexmela) corona* (Gmelin, 1791), length 78 mm, Rabbit Key, Ten Thousand Islands, Florida; **C, D=** *Melongena (Rexmela) corona* (Gmelin, 1791), length 73 mm, Kice Island, Ten Thousand Islands, Florida; **E, F=** *Melongena (Rexmela) corona* (Gmelin, 1791), length 49 mm, Camp Lulu Key, Ten Thousand Islands, Florida.

Figure 4.8 ***Melongena (Rexmela) corona*** **variants** ***belknapi*** **and** ***trinodulosa.***
A, B= *Melongena (Rexmela) corona* form *belknapi* Petit de la Saussaye, 1852, length 43 mm, Turtle Key, Ten Thousand Islands; **C, D=** *Melongena (Rexmela) corona* form *trinodulosa* Emery and Lermond, 1936, length 62 mm, Jewell Key, Ten Thousand Islands; **E, F=** *Melongena (Rexmela) corona* form with webbed varices, length 51 mm, Chokoloskee Island, Ten Thousand Islands.

Figure 4.9 ***Melongena (Rexmela) corona*** **form** ***perspinosa.***
A, B= *Melongena (Rexmela) corona* form *perspinosa* Pilsbry and Vanatta, 1934, length 65 mm; **C, D=** *Melongena (Rexmela) corona* form *perspinosa* Pilsbry and Vanatta, 1934, white color form, length 56 mm; **E, F=** *Melongena (Rexmela) corona* form *perspinosa* Pilsbry and Vanatta, 1934, length 60 mm. All specimens from Chokoloskee Island, Ten Thousand Islands, Collier County, Florida.

Figure 4.10 *Fulguropsis pyruloides* from the Suwannean Subprovince.
A, B= *Fulguropsis pyruloides* (Say, 1822), length 77 mm, Rabbit Key, Ten Thousand Islands; **C, D**= *Fulguropsis pyruloides* (Say, 1822), length 68 mm, Turtle Key, Ten Thousand Islands; **E, F**= *Fulguropsis pyruloides* (Say, 1822), golden-orange color form, length 62 mm, Rabbit Key, Ten Thousand Islands. This smooth, silky-textured western Florida species occurs together with the heavily-ribbed *Fulguropsis keysensis* only along the Ten Thousand Islands.

Figure 4.11 ***Americoliva sayana sarasotaensis*** **from the Suwannean Subprovince.**
A, B= *Americoliva sayana sarasotaensis* (Petuch and Sargent, 1986), typical color form, length 47 mm; **C, D=** *Americoliva sayana sarasotaensis* (Petuch and Sargent, 1986), pale color form, length mm, 44 mm; **E, F=** *Americoliva sayana sarasotaensis* (Petuch and Sargent, 1986), golden color form, length 32 mm. All specimens from Kice Island, Ten Thousand Islands, Collier County, Florida. The nominate subspecies, *Americoliva sayana sayana* (Ravenel, 1834), is restricted to the southeastern US coast, from North Carolina to Boca Raton, Florida and is absent from Biscayne Bay, Miami, and the Florida Keys. In the Ten Thousand Islands and Suwannean Subprovince area of western Florida, the nominate subspecies is replaced by the subspecies *sarasotaensis*, which differs in being a smaller, stockier, and more inflated shell.

Figure 4.12 *Gradiconus floridanus* from the Suwannean Subprovince.
A, B= *Gradiconus floridanus* (Gabb, 1869), length 41 mm; **C, D**= *Gradiconus floridanus* (Gabb, 1869), length 42 mm; **E, F**= *Gradiconus floridanus* (Gabb, 1869), length 39 mm. All specimens from Kice Island, Ten Thousand Islands, Collier County, Florida. Erroneously called "*G. anabathrum*" for years, but that taxon was found to be a synonym of the Panamic species *G. scalaris* (Berschauer, 2022). Also incorrectly stated, in several recent publications, as being found along Texas and on the Flower Garden Reefs; but those shells were later found to be misidentified *G. largillierti* and true *G. floridanus* is now known to be restricted to southwest Florida.

Figure 4.13 *Gradiconus floridanus* form *floridensis*.
A, B= *Gradiconus floridanus* form *floridensis* (Sowerby II, 1870), length 37 mm; **C, D**= *Gradiconus floridanus* form *floridensis* (Sowerby II, 1870), length 47 mm; **E, F**= *Gradiconus floridanus* form *floridensis* (Sowerby II, 1870), length 42 mm. All specimens are from Camp Lulu Key, Ten Thousand Islands, Collier County, Florida.

Figure 4.14 ***Jaspidiconus stearnsi*** **from the Suwannean Subprovince.**
A, B= *Jaspidiconus stearnsi* (Conrad, 1869), length 23 mm, Goodland Bay, Collier County, Florida; **C, D=** *Jaspidiconus stearnsi* (Conrad, 1869), lilac color form, length 21 mm, Camp Lulu Key, Ten Thousand Islands, Florida; **E, F=** *Jaspidiconus stearnsi* (Conrad, 1869), length 20 mm, Kice Island, Ten Thousand Islands, Collier County, Florida.

Figure 4.15 ***Argopecten irradians taylorae*** **from the Floridian and Suwannean Subprovinces.**
A, B= *Argopecten irradians taylorae* Petuch, 1987, length 34 mm, Joe Kemp Key, Florida Bay, Florida; **C, D=** *Argopecten irradians taylorae* Petuch, 1987, orange color form, length 36 mm, Joe Kemp Key, Florida Bay, Florida; **E=** *Argopecten irradians taylorae* Petuch, 1987, red color form, length 26 mm, Kice Island, Ten Thousand Islands; **F=** *Argopecten irradians taylorae* Petuch, 1987, large specimen, length 57 mm, Goodland Bay, Collier County, Florida.

Figure 4.16 Deep Neritic Species of the Suwannean Subprovince.
A, B, C= *Xenophora microdiscus* Petuch, 1994, diameter 17 mm, dredged from 150 m depth west of Tampa, Florida (originally described as a late Pleistocene Florida fossil, but now found to be living in deep water in the Gulf of Mexico); **D, E=** *Cyphoma lindae* Petuch, 1994, length 22 mm, dredged from 220 m depth west of Tampa, Florida (mantle of live animal has small black dots); **F, G=** *Cancellaria richardpetiti* Petuch, 1987, length 47 mm, from 200 m depth south of Apalachicola, Florida; **H, I=** *Conomodulus lambi* Petuch and Berschauer, 2023, holotype, diameter 13 mm, trawled from 200 m depth north of the Dry Tortugas, Florida Keys.

Figure 4.17 Deep Water Endemic Muricids of the Suwannean Subprovince.
A, B= *Favartia lindae* Petuch, 1987, length 18 mm, from 100 m depth west of Tampa, Florida; **C=** *Favartia lindae* Petuch, 1987, holotype, length 11 mm, dredged from 150 m depth, south of Apalachicola, Florida; **D, E=** *Vokesimurex lindajoyceae* Petuch, 1987, length 32.4 mm, from 200 m depth west of Tampa, Florida; **F, G=** *Vokesimurex lindajoyceae* Petuch, 1987, length 29 mm, 200 m depth off Cedar Key, Florida (incorrectly referred to as "*Murex anniae*" by many workers; but that species is actually an early Pleistocene Florida fossil); **H, I=** *Murexiella taylorae* Petuch, 1987, length 15 mm, from 100 m depth off Tampa, Florida; **J=** *Murexiella taylorae* Petuch, 1987, holotype, length 16 mm, from 200 m depth off Cedar Key, Levy County, Florida.

Figure 4.18 Endemic Deep Water Species of the Suwannean Subprovince.
A, B= *Heilprinia lindae* Petuch, 1987, holotype, length 143 mm, from 150 m depth south of Apalachicola, Florida (note the deeply-indented thin channel around the sutures of the spire whorls, which is absent on the similar shallower-water *H. timessus* shown in Figure 1.4 in Chapter 1); **C, D**= *Hesperisternia harasewychi* (Petuch, 1987), holotype, length 25 mm, from 150 m depth south of Apalachicola, Florida; **E**= *Torculoidella lindae* Petuch, 1987, holotype, length 22.5 mm, dredged from 150 m depth south of Apalachicola, Florida; **F, G**= *Hesperisternia sulzyckii* Petuch and Myers, 2014, holotype, length 23.6 mm, 50 m depth north of the Dry Tortugas, Florida Keys; **H**= *Aristofusus stegeri* (Lyons, 1978), length 62 mm, 150 m depth west of Tampa, Florida.

Figure 4.19 *Clenchina marionae* from the Suwannean Bathyal Areas.
A, B= *Clenchina marionae* (Pilsbry and Olsson, 1953), length 48 mm, 400 m depth NW of the Dry Tortugas; **C, D=** *Clenchina marionae* (Pilsbry and Olsson, 1953), holotype, length 59 mm, 146 m depth west of Tampa, Florida.

Figure 4.20 Endemic Deep Water Species of the Suwannean Subprovince.
A, B= *Americoliva sunderlandi* (Petuch, 1987), length 21 mm, from 150 m depth west of Cedar Key, Levy County, Florida; **C, D=** *Americoliva sunderlandi* (Petuch, 1987), holotype, length 22 mm, from 150 m depth west of Cedar Key, Levy County, Florida; **E=** *Cinguloterebra lindae* (Petuch, 1987), length 68 mm, from 200 m depth south of Apalachicola, Florida; **F, G=** *Cinguloterebra lindae* (Petuch, 1987), holotype, length 63 mm, from 150 m depth south of Apalachicola, Florida (differs from the widespread *C. floridana* in being much more slender and in having large brown patches on the whorls); **H, I=** *Polystira sunderlandi* Petuch, 1987, holotype, length 28 mm, from 150 m depth south of Apalachicola, Florida; **J=** *Fenimorea brunnescens* (Rehder, 1943), length 15 mm, from 100 m depth north of the Dry Tortugas, Florida Keys.

Figure 4.21 Endemic Deep Water Suwannean Cone Shells.
A, B= *Dauciconus aureonimbosus* (Petuch, 1987), holotype, length 27 mm, from 150 m depth south of Apalachicola, Florida (also dredged off Alabama and Louisiana); **C, D=** *Dauciconus aureonimbosus* (Petuch, 1987), dark orange color form, length 26 mm, from 200 m depth north of the Dry Tortugas, Florida Keys; **E, F=** *Gradiconus optabilis* (A. Adams, 1854), length 25.4 mm, from 100 m depth north of the Dry Tortugas, Florida Keys; **G, H=** *Gradiconus optabilis* (A. Adams, 1854), holotype, length 22 mm, type locality unknown; now presumed to be the Gulf of Mexico off Florida. Freshly-collected specimens, like E and F shown here, have a bright pinkish-rose color within the aperture.

Figure 4.22 *Chicoreus rachelcarsonae* from the West Florida Coralline Algal Beds.
A, B= *Chicoreus rachelcarsonae* Petuch, 1987, length 40 mm, from 200 m depth on coralline algal beds, west of Tampa, Florida; **C, D=** *Chicoreus rachelcarsonae* Petuch, 1987, holotype, length 43 mm, from 150 m depth south of Apalachicola, Florida; also dredged off the Alabama coast. This species always has 5 frondose spines of each varix, as opposed to 6 spines on the congeneric *C. dilectus*.

Figure 4.23 *Phyllonotus whymani* from the West Florida Coralline Algal Beds.
A, B= *Phyllonotus whymani* Petuch and Sargent, 2011, holotype, length 43.4 mm; **C, D**= *Phyllonotus whymani* Petuch and Sargent, 2011, length 45 mm; **E, F**= *Phyllonotus whymani* Petuch and Sargent, 2011, length 67.9 mm. All three specimens from 35 m depth north of the Dry Tortugas, Florida Keys (original description reprinted in the Systematic Appendix at the end of this book).

Figure 4.24 *Lindafulgur lyonsi* from the West Florida Coralline Algal Beds.
A, B, C= *Lindafulgur lyonsi* (Petuch, 1987), holotype, length 132 mm, dredged from 170 m depth west of Clearwater, Florida, during the "Hourglass" marine survey project, Florida Department of Natural Resources, in 1979. This rare and seldom-seen busyconid lives together with the slit shell, *Perotrochus amabilis* (F. Bayer, 1963), and the abalone, *Haliotis pourtalesii* Dall, 1881, on the deep water sponge bioherms and coralline algal beds. This is the only-known record of a busyconid whelk, an abalone, and a slit shell living together in the same ecosystem. *Lindafulgur lyonsi* is similar in appearance to *L. camazotz* Petuch and Berschauer, 2023 from the Yucatanean Subprovince (Figure 6.27, Chapter 6), but it differs in being a much thinner and more delicate shell that lacks strong spiral cords on the body whorl and spire whorls.

Figure 4.25 *Lindapecten lindae* from the West Florida Coralline Algal Beds.
A, B= *Lindapecten lindae* Petuch, 1995, length 22 mm, from 200 m depth on coralline algal bed, west of Tampa, Florida; **C, D=** *Lindapecten lindae* Petuch, 1995, length 18.8 mm, from 200 m depth on coralline algal bed, north of the Dry Tortugas, Florida Keys. Differs from the widespread shallow water *Lindapecten muscosus* (Wood, 1828) in being much smaller, in having distinctly flatter valves, and in having proportionally-larger spines on the ribs.

Figure 4.26 *Petaloconchus nigricans* aggregation from a Chokoloskean Vermetoherm.
Close-up of a fragment, length 310 mm, from an aggregation of the vermetid gastropod, *Petaloconchus nigricans* (Dall, 1884), collected from the beach adjacent to a worm shell "reef" along Turtle Key, Ten Thousand Islands. This unusual gregarious vermetid forms immense reef-like structures (vermetoherms) that extend for acres along some of the outer Ten Thousand Islands adjacent to Everglades National Park (see Petuch and Berschauer, 2022).

Figure 4.27 Endemic Taxa of the Chokoloskean Infraprovince.
A= *Vermicularia fargoi owensi* Petuch and Myers, 2014, length 130 mm, from Pavilion Key, Ten Thousand Islands, Florida; **B, C=** *Vermicularia fargoi owensi* Petuch and Myers, 2014, holotype (juvenile), length 23 mm, Turtle Key, Ten Thousand Islands, Florida (originally incorrectly listed as coming from Demijohn Key); **D, E=** *Gemophos tinctus pacei* Petuch and Sargent, 2011, length 31 mm. Goodland Bay, Collier County, Florida (original description reprinted in the Systematic Appendix at the end of this book); **F=** *Gemophos tinctus pacei* Petuch and Sargent, 2011, length 27 mm, Goodland Bay, Collier County, Florida; **G, H=** *Naticarius verae* Rehder, 1947, length 32 mm, Kice Island, Ten Thousand Islands, Florida; **I=** *Naticarius verae* Rehder, 1947, length 28 mm, Kice Island, Ten Thousand Islands, Florida.

Figure 4.28 *Cinctura tortugana foxi* from the Chokoloskean Infraprovince, the western Florida Inner Shelf, and the Florida Panhandle.

A, B= *Cinctura tortugana foxi* Petuch and Berschauer, 2022, length 83 mm, Camp Lulu Key, Ten Thousand Islands, Florida; **C, D=** *Cinctura tortugana foxi* Petuch and Berschauer, 2022, length 86 mm, Kice Island, Ten Thousand Islands, Florida; **E, F=** *Cinctura tortugana foxi* Petuch and Berschauer, 2022, length 92.9 mm, red color variant from off Panama City, Bay County, Florida. Although most common in the Ten Thousand Islands, the subspecies *foxi* ranges sporadically along western Florida in shallow subtidal areas. Northern populations frequently are red or solid orange.

Figure 4.29 ***Mercenaria browni*** **from the Chokoloskean Infraprovince.**
A, B= *Mercenaria browni* Petuch and Berschauer, 2019, heavily-sculptured juvenile, width 49 mm; **C=** *Mercenaria browni* Petuch and Berschauer, 2019, adult specimen, width 91 mm; **D=** *Mercenaria browni* Petuch and Berschauer, 2019, ornamented juvenile specimen, width 41 mm. All specimens from the Rabbit Key Pass, adjacent to Turtle Key, Ten Thousand Islands, Florida.

Figure 4.30 Endemic Taxa of the Apalachicolan Infraprovince.
A, B= *Cinctura hunteria apalachee* Petuch and Berschauer, 2020, length 64 mm, 1 m depth in sea grasses, Sandy Point, St. Andrew Bay, Bay County, Florida; **C, D=** *Cinctura hunteria apalachee* Petuch and Berschauer, 2020, form with multiple bands, length 79 mm, 1 m depth in sea grasses, Sandy Point, St. Andrew Bay, Bay County, Florida; **E, F=** *Hesperisternia grandana* (Abbott, 1986), length 27 mm, in Turtle Grass, 1 m depth off Panacea, Florida; endemic to the Big Bend Sea Grass Preserve area; **G, H=** *Modulus danielsi* Petuch and Berschauer, 2023, holotype, width 10 mm, 1 m depth in sea grasses, Sandy Point, St. Andrew Bay, Bay County, Florida.

Figure 4.31 ***Melongena (Rexmela) corona johnstonei*** **from the Apalachicolan Infraprovince.**
A= *Melongena (Rexmela) corona johnstonei* Clench and Turner, 1956, length 130 mm, Destin Harbor, Destin, Okaloosa County, Florida; **B, C=** *Melongena (Rexmela) corona johnstonei* Clench and Turner, 1956, length 116 mm, in Dickerson Bay, Panacea, Wakulla County, Florida; **D, E=** *Melongena (Rexmela) corona johnstonei* Clench and Turner, 1956, length 105 mm, in Deadman Bay, Steinhatchee, Taylor County, Florida; **F, G=** *Melongena (Rexmela) corona johnstonei* form *incurvata* Lermond, 1936, length 104 mm, Cedar Key, Levy County, Florida.

Figure 4.32 ***Americoliva bollingi choctaw*** **from the Apalachicolan Infraprovince.**
A, B= *Americoliva bollingi choctaw* Petuch and Myers, 2014, length 44 mm; **C, D=** *Americoliva bollingi choctaw* Petuch and Myers, 2014, length 42 mm; **E, F=** *Americoliva bollingi choctaw* Petuch and Myers, 2014, length 39 mm. All specimens were found at Sandy Point, St. Andrew Bay, Bay County, Florida.

Figure 4.33 The Apalachicolan Infraprovince *Scaphella johnstoneae*.
A, B= *Scaphella johnstoneae* Clench, 1953, length 119 mm, from 30 m depth off Apalachicola, Florida; **C, D=** *Scaphella johnstoneae* Clench, 1953, paratype, length 83.5 mm, 20 m depth off Dauphin Island, Alabama; **E=** *Scaphella johnstoneae* Clench, 1953, length 126 mm, 25 m depth off Mobile, Alabama; **F=** *Scaphella johnstoneae* Clench, 1953, length 102 mm, 35 m depth south of Panacea, Big Bend area of northwestern Florida.

Figure 4.34 *Jaspidiconus forticostatus* from the Big Bend of northwestern Florida.
A, B= *Jaspidiconus forticostatus* Petuch and Berschauer, 2024, holotype, length 25.8 mm; **C, D=** *Jaspidiconus forticostatus* Petuch and Berschauer, 2024, length 26.7 mm; **E, F=** *Jaspidiconus forticostatus* Petuch and Berschauer, 2024, length 22.7 mm. All three specimens were dredged from 50 m depth due west of Cedar Key, Levy County, Florida, in the Florida Big Bend embayment.

CHAPTER 5. Endemic Mollusks of the Texan Subprovince

Named for the State of Texas (Petuch, 2013; Petuch and Berschauer, 2021a), the Texan Subprovince extends from the western side of the Mississippi River Delta, Louisiana to Veracruz, Mexico and encompasses the entire Texan coastline and continental shelf and the western side of the Bay of Campeche (Figure 5.1). Compared to the other three Carolinian subprovinces present within the Gulf of Mexico, the Texan Subprovince has the most impoverished macromolluscan fauna, with the lowest biodiversity. This faunal impoverishment reflects the harsh environmental conditions of the Texas and northeastern Mexican coasts, where air temperatures vary seasonally; from below freezing during the winter to reaching high tropical conditions during the summer. The coastal lagoons that extend along much of the coastline of the Texan Subprovince typically have wildly fluctuating salinities, depending on the season; with hyposaline conditions (10-20 ppt salinity) in the summer, during the heavy monsoonal rains, and with hypersaline conditions (40-50 ppt salinity) during the dry, windy winter months. The combination of the extreme air and water temperatures, together with the wildly fluctuating salinities, has resulted in a scaled-down molluscan fauna with its own distinctive composition. A detailed overview of the Texan Subprovince coastal environments and their associated molluscan assemblages was given by Tunnell, Andrews, Barrera, and Moretzsohn in their comprehensive 2010 book, "*Encyclopedia of Texas Seashells*". This large text is also an important source of information on the taxonomy and biodiversity of the myriad of micromollusks that inhabit the Texan coastal lagoons and shallow continental shelf. For a detailed overview of the molluscan fauna of the West Flower Garden Reef southeast of Galveston, and its associated coral banks, see Bright and Pequegnat (1974).

Endemic Mollusks of the Coastal Lagoons

Besides supporting the typical array of Carolinian Province shallow water mollusks, such as the bivalves *Mercenaria campechiensis* and *Dinocardium vanhyningi* and the gastropods *Strombus alatus* and *Triplofusus giganteus*, the Texas coastal lagoons also house a number of interesting endemic taxa. Within the sheltered bays and lagoons, particularly in areas where intertidal banks of muddy sand have formed, the small endemic Texas Venus Clam, *Mercenaria texana* (Figure 5.5), often forms large beds and creates the trophic base for an entire endemic ecosystem. The large busyconid whelk, *Sinistrofulgur pulleyi* (Figure 5.8; the Texas State Shell), is the primary predator on these lagoonal clam beds, where it also feeds on the larger sympatric congener, *Mercenaria campechiensis*. A similar analogue ecosystem is seen in the lagoons of the Ten Thousand Islands of southwestern Florida, in the Suwannean Subprovince (Petuch and Berschauer, 2022). Here the small endemic lagoonal venerid, *Mercenaria browni*, lives along with the larger, widespread *Mercenaria campechiensis* and both species are the primary food resource for the large busyconid *Sinistrofulgur sinistrum*. Another similar example is seen in the coastal lagoons of Palm Beach County, Florida where the small venerid

Mercenaria hartae lives together with the larger *Mercenaria campechiensis* and both species are frequently preyed upon by the large busyconid *Sinistrofulgur sinistrum* (see Petuch and Myers, 2014 for illustrations of *M. hartae*).

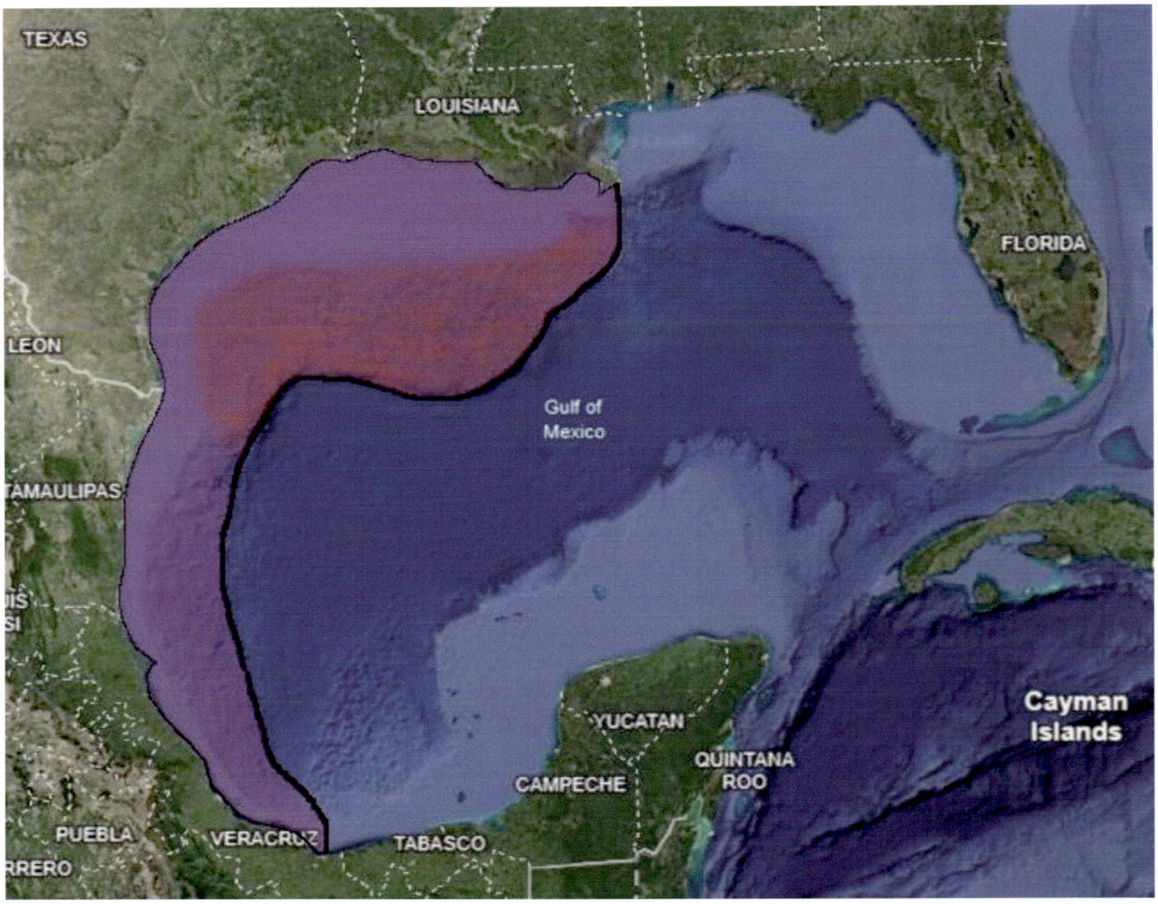

Figure 5.1 Map showing the geographical extent of the Texan Subprovince (purple and rose shading) of the Carolinian Molluscan Province. The area containing the best-developed petroleum seeps and brine pools is designated by the rose shading.

Figure 5.2 Aerial view of a small sandy islet and the adjacent oyster banks, along the First Chain of Islands, between San Antonio Bay and Espirito Santo Bay, on the western side of Matagorda Island, Texas. The long, undulating bank, composed entirely of *Crassostrea virginica* and *Ostreola equestris* oysters, acts as a barrier to tidal flow between the two bays and supports numerous endemic predatory gastropods such as the fasciolariid *Cinctura lilium*, the busyconid *Sinistrofulgur pulleyi*, and the muricid *Stramonita alderi*. Extensive oyster banks like this are characteristic of the lagoon systems along the Texas coast (photo courtesy of Scott Sommerlatte).

The Cordgrass (*Spartina*) meadows and mud flats of the inner protected bays are the primary habitat for two species of potamidid gastropods, the small *Cerithidea hegewischii* (Figure 5.4), which is endemic to the Texan Subprovince salt marshlands, and *Cerithidea pliculosa*, a larger species that is widespread throughout the western Gulf of Mexico and Caribbean Sea. These muddy areas also support large banks of the oysters *Ostreola equestris* and *Crassostrea virginica* (Figure 5.2) which serve as prey items for several endemic predatory gastropods, including the large eurytopic busyconid *Sinistrofulgur pulleyi*, the rare muricid *Stramonita alderi* (Figure 5.4), and the fasciolariid *Cinctura lilium*. Often, within these quiet protected lagoons, large beds of the sea grasses *Thalassia* and *Halodule* will develop and cover acres of the shallow sea floor. These protected, quiet water sea grass beds support immense shoals of the Bay Scallop, *Argopecten irradians amplicostatus* (Figure 5.4), a distinct subspecies (or sibling species) of pectinid that is restricted to the Texan Subprovince. Intertidal sand flats, near areas of higher salinities and open oceanic water conditions, typically house large populations of the endemic Texan Olive Shell, *Americoliva sayana texana* (Figure 5.6). This predatory volutoidean gastropod represents the western Gulf subspecies of a taxonomic complex that also includes the eastern Gulf *A. sayana sarasotaensis* (see Chapter 4) and the eastern United States *A. sayana sayana* (see Petuch and Sargent, 1986 for details on differences of shell morphology). No members of this complex occur in the Floridian Subprovince and the two Gulf subspecies are genetically isolated.

Endemic Mollusks of the Texas Neritic Environments

The neritic environment of the Texas continental shelf, from 10 m to 100 m depths, supports a large number of endemic mollusks, including some of the most beautiful and interesting found in the entire Gulf of Mexico. The wide shallow shelf also houses a large fauna of classic Carolinian Province index species, including the pectinid bivalve *Nodipecten fragosus*, the large muricid *Hexaplex fulvescens*, and the large cypraeid *Macrocypraea (Lorenzicypraea) cervus*, and also widespread western Atlantic species such as the epitoniid *Amaea mitchelli*. Living along with these large and conspicuous mollusks are a number of characteristic Texan endemic species, including the naticid *Neverita duplicata texasiana* (Figure 5.15), the busyconids *Fulguropsis texanus* (Figure 5.9) and *Fulguropsis galvestonense* (Figure 5.10), the gorgonian-associated (*Leptogorgia*) ovulid *Simnialena marferula* (Figure 5.4), and the strikingly-patterned volute *Scaphella curryi* (Figures 5.12 and 5.13). This latter volute species is now known to have two main populations: a southern population that ranges along the eastern coast of Mexico and into the Bay of Campeche; and a northern population found along the Texas coast. The Bay of Campeche form has a stockier shell with a more inflated body whorl and a color pattern composed of large dark red checkers that often fuse into wide bands. The Texan form has a narrower, more elongated body whorl with a proportionally-higher spire and a color pattern composed of more numerous and proportionally-smaller dark red checkers. Both forms of this characteristic Texan Subprovince volute are shown here in Figure 5.13.

At the outer edge of the northern section of the Texas continental shelf, in depths averaging 150-200 m, another endemic Texas Subprovince scaphelline volute occurs, the large *Aurinia dubia ethelae* (Figure 5.11). This highly-inflated and fragile deep water volute has a distinctive sculpture pattern on the body whorl, producing large, widely-spaced fold-like ribs.

Figure 5.3 A pair of the endangered American Oystercatcher (*Haemotopus palliatus* Temminck, 1820) searching for food on piles of shell hash along the First Chain of Islands, between San Antonio Bay and Espirito Santo Bay, on the western side of Matagorda Island, Texas. Feeding on mudflats and sandbars at low tide, this colorful shorebird feeds almost exclusively on bivalved mollusks and is one of the principal predators of the endemic intertidal Texas venerid, *Mercenaria texana* (photo courtesy of Scott Sommerlatte).

This same depth range houses two other widespread large scaphelline volutes, the heavily-sculptured *Clenchina robusta* and *Aurinia kieneri*, one of the largest volute species found in the Gulf of Mexico (both shown in Chapter 2). Living along with this outer neritic volute fauna is *Dalliconus sauros* (illustrated in Chapter 2, Figure 2.17), a small, heavily-sculptured cone shell that ranges along the Texas shelf, eastern Mexican coast, and off the Yucatan Peninsula. Along the escarpment at the far-eastern edge of the Texas shelf, numerous petroleum seeps and brine pools occur and these support massive methanophilic bacteria growth that is the ecological foundation for the bizarre giant mussel-based communities that are discussed in Chapter 2. One of the rarest and least-known Texas volutes, the strange, turreted *Clenchina worki* (see Chapter 2), lives in association with these petroleum seep ecosystems. The coral-covered knolls along the edge of the Texas continental shelf, collectively referred to as the Flower Garden Reefs, are now known to house an odd, impoverished mixture of Caribbean and Carolinian malacofaunas, unlike any others found along the Texas coast or northern Gulf of Mexico. A recently-described conid gastropod, *Jaspidiconus hortusensis* (Figure 5.14), has been collected on West Flower Garden Reef and appears to be endemic to that reef complex. See Bright and Pequegnat (1974) for an overview of the molluscan fauna of West Flower Garden Reef.

TABLE 6. Important Mollusks that are Endemic to the Texan Subprovince.

Gastropoda
Potamididae
Cerithidea hegewischii (Philippi, 1848)
Ovulidae
Simnialena marferula Cate, 1973
Naticidate
Neverita (Glossaulax) duplicata texasiana (Philippi, 1849)
Muricidae
Stramonita alderi Petuch and Berschauer, 2020
Fasciolariidae
Cinctura lilium (Fischer von Waldheim, 1807)
Busyconidae
Fulguropsis galvestonense (Hollister, 1958)
Fulguropsis texanus (Hollister, 1958)
Sinistrofulgur pulleyi (Hollister, 1958)
Volutidae
Aurinia dubia ethelae Pilsbry and Olsson, 1953 (Mississippi River Delta and NE Texas)
Scaphella curryi Petuch and Berschauer, 2019
Olividae
Americoliva sayana texana (Petuch and Sargent, 1986)
Conidae
Jaspidiconus hortusensis Petuch and Berschauer, 2024 (endemic to the Flower Garden Reefs)
Bivalvia
Pectinidae
Argopecten irradians amplicostatus (Dall, 1898)
Veneridae
Mercenaria texana (Dall, 1902)

ICONOGRAPHY OF TEXAN SUBPROVINCE ENDEMIC SPECIES

Figure 5.4 Endemic Mollusks of the Texan Subprovince.
A, B= *Cerithidea hegewischii* (Philippi, 1848), length 15 mm, on mud flats, Aransas Pass, Port Aransas, Texas; **C, D=** *Stramonita alderi* Petuch and Berschauer, 2020, holotype, length 35 mm, Fish Pass Jetty, Corpus Christi, Texas (identified as "*Stramonita* cf. *rustica*" by Tunnell *et al.*, 2010: 212); **E, F=** *Argopecten irradians amplicostatus* (Dall, 1898), length 33 mm, 1 m depth off Mustang island, Corpus Christi, Nueces County, Texas; **G, H=** *Simnialena marferula* Cate, 1973, length 11 mm, on *Leptogorgia* gorgonian washed onto beach at San Jose Island, Port Aransas, Texas.

Figure 5.5 *Mercenaria texana* from the Texas Coastal Lagoons.
A, B= *Mercenaria texana* (Dall, 1902), width 63 mm, Port Lavaca, Matagorda Bay, Matagorda County, Texas; **C, D=** *Mercenaria texana* (Dall, 1902), width 39 mm, Espiritu Santo Bay, Matagorda Island, Texas; **E=** *Mercenaria texana* (Dall, 1902), width 34 mm, East Matagorda Bay, Matagorda Beach, Texas. This endemic Texan intertidal sandflat bivalve is one of the principal food resources for the American Oystercatcher, *Haematopus palliatus* (see Figure 2.3).

Figure 5.6 *Americoliva sayana texana* from the Texan Coastal Lagoons.
A, B= *Americoliva sayana texana* (Petuch and Sargent, 1986), length 52 mm, on sand flat at low tide, Aransas Pass, Port Aransas, Texas; **C, D=** *Americoliva sayana texana* (Petuch and Sargent, 1986), length 48 mm, Aransas Pass, Port Aransas, Texas; **E, F=** *Americoliva sayana texana* (Petuch and Sargent, 1986), length 46 mm, Laguna Madre, South Padre Island, Texas.

Figure 5.7 *Cinctura lilium* from the Texan Subprovince.
A, B= *Cinctura lilium* (Fischer von Waldheim, 1807), length 100 mm, low tide, San Jose Island, Port Aransas, Texas; **C, D=** *Cinctura lilium* (Fischer von Waldheim, 1807), length 96 mm, low tide, San Jose Island, Port Aransas, Texas; **E, F=** *Cinctura lilium* (Fischer von Waldheim, 1807), length 122.4 mm, washed onto beach, South Padre Island, Texas.

Figure 5.8 ***Sinistrofulgur pulleyi*** **from the Texan Subprovince.**
A, B= *Sinistrofulgur pulleyi* (Hollister, 1958), length 154 mm, at low tide in West Bay, Galveston, Texas; **C, D=** *Sinistrofulgur pulleyi* (Hollister, 1958), length 118 mm, at low tide in West Bay, Galveston, Texas; **E, F=** *Sinistrofulgur pulleyi* (Hollister, 1958), length 109 mm, Matagorda Beach, Matagorda, Texas. This is the official State Shell of Texas.

Figure 5.9 *Fulguropsis texanus* from the Texan Subprovince.
A, B= *Fulguropsis texanus* (Hollister, 1958), heavily-ribbed specimen, length 90 mm, exposed at low tide, West Bay, Galveston Island, Galveston, Texas; **C, D=** *Fulguropsis texanus* (Hollister, 1958), length 113 mm, washed onto beach, San Jose Island, Port Aransas, Texas. Also found in the Bay of Campeche and the adjacent Campeche Bank.

Figurε 5.10 *Fulguropsis galvestonense* from the Texan Subprovince.
A, B= *Fulguropsis galvestonense* (Hollister, 1958), specimen with the classic sloped and carinated shoulder, length 114 mm, tawled from 10 m depth off Anahuac, Bolivar Peninsula, Texas; **C, D=** *Fulguropsis galvestonense* (Hollister, 1958), length 65 mm, 10 m depth off Freeport, Texas. Also found on the Campeche Bank and adjacent areas of the Yucatεn Peninsula.

Figure 5.11 ***Aurinia dubia ethelae*** **from the Mississippi Delta and northern Texas.**
A, B= *Aurinia dubia ethelae* Pilsbry and Olsson, 1953, length 162 mm, dredged from 250 m depth, 60 km SW of Atchafalaya Bay, western Mississippi River Delta, Louisiana; **C, D=** *Aurinia dubia ethelae* Pilsbry and Olsson, 1953, length 136.0 mm, dredged from 250 m depth east of Galveston, Texas.

Figure 5.12 *Scaphella curryi* from Texas and Eastern Mexico.
A, B= *Scaphella curryi* Petuch and Berschauer, 2019, length 107.8 mm; **C, D=** *Scaphella curryi* Petuch and Berschauer, 2019, length 106.0 mm; **E, F=** *Scaphella curryi* Petuch and Berschauer, 2019, holotype, length 101.2 mm. All three specimens were trawled by shrimpers from 100 m depth east of Veracruz, Veracruz State, Mexico.

Figure 5.13 *Scaphella curryi* Variants.
A, B= *Scaphella curryi* Petuch and Berschauer, 2019, banded color form, length 99.5 mm, from 100 m depth in the Bay of Campeche, east of Veracruz, Mexico; **C, D=** *Scaphella curryi* Petuch and Berschauer, 2019, length 98 mm. trawled from 100 m depth in the Bay of Campeche, east of Veracruz, Mexico; **E, F=** *Scaphella curryi* Petuch and Berschauer, 2019, elongated and high-spired, possible undescribed northern subspecies, length 113 mm, trawled by shrimpers from 100 m depth off the Aransas Pass, Port Aransas, Nueces County, Texas.

Figure 5.14 Cone Shells from West Flower Garden Reef, Texas.
A, B= *Jaspidiconus hortusensis* Petuch and Berschauer, 2024, holotype length 17.1 mm; **C=** *Jaspidiconus hortusensis* Petuch and Berschauer, 2024, paratype, length 22.5 mm (in the Biodiversity Research and Teaching Collection (BRTC), Texas A & M University). Both specimens were collected in 25-27 m depth, on carbonate sand, West Flower Garden Reef, 160 km SE of Galveston, Texas (Dr. Thomas J. Bright, collector). This small cone is the only endemic mollusk known from West Flower Garden Reef and is the only *Jaspidiconus* species known from the Texan Subprovince; **D, E=** *Chelyconus testudinarius* (Hwass, 1792), solid yellow-orange color form, under coral slabs, 10 m depth on West Flower Garden Reef, Texas. *Chelyconus testudinarius* from the northern Gulf reefs, banks, and oil platforms, are generally solid orange in color.

Figure 5.15 The Texan endemic Naticid, *Neverita (Glossaulax) duplicata texasiana* (Philippi, 1849).
A, B= *Neverita (Glossaulax) duplicata texasiana* (Philippi, 1849), width 45 mm; **C, D=** *Neverita (Glossaulax) duplicata texasiana* (Philippi, 1849), with 34 mm; **E=** *Neverita (Glossaulax) duplicata texasiana* (Philippi, 1849), base of the 34 mm specimen, showing the large umbilical callus which almost completely covers the umbilical area, leaving a narrow slot along the anterior side. This isolated Texas subspecies differs from the northern and eastern Gulf *Neverita (Glossaulax) duplicata campechiensis* in being a much more flattened and discoidal shell with a much larger and better-developed umbilical callus (see Figure 2.18, Chapter 2). Both specimens are from Matagorda Bay, Matagorda County, Texas.

CHAPTER 6. Endemic Mollusks of the Yucatanean Subprovince

Extending from approximately Coatzacoalcos, Tabasco State, to Isla Cozumel, Quintana Roo State, and including the States of Campeche and Yucatan, the Yucatanean Subprovince is the most tropical of the four Gulf Carolinian Subprovinces (Petuch, 2013; Petuch and Berschauer, 2021a). Water temperatures around the peninsula rarely fall below 21 degrees C (70 degrees F) and living coral reefs abound in most offshore areas. Because of the great distance from the quartz sand beaches of the Texan Subprovince, the Yucatanean sediments are dominated by carbonates, being composed primarily of crushed shells, foraminiferal tests, micromollusks, coralline algae phytoliths, calcareous sponge spicules, and chemogenic precipitates. Because of these oceanographic and ecological aspects, the Yucatanean Subprovince is a close analogue of the Floridian Subprovince, with both subprovincial areas sharing similar coral reef tracts and carbonate sediment environments. Since the entire Yucatan Peninsula is much farther south than the Florida Keys, and is under the climatic influence of warmer Caribbean air masses and water currents, the Yucatanean Subprovince never experiences the extreme cold spells that affect Florida and the Floridian Subprovince during the winter.

The area of the Yucatanean Subprovince can be further subdivided into two distinct ecoregions; one that extends from Coatzacoalcos, Tabasco State to Rio Lagartos, Yucatan State, and includes the entire **Bay of Campeche** and the wide **Campeche Bank** and its coral islands; and another that extends from Rio Lagartos, across the Laguna Yalahau and Holbox area, and south to Isla Mujeres, Quintana Roo State. Some Yucatanean faunal elements also extend south to the Playa del Carmen and Cozumel Island area, where they mix with the predominantly-Caribbean Province faunas. Across the entire subprovince, a number of widespread characteristic gastropod species occur in every offshore environment and coastal area. The most conspicuous of these include the muricid *Vokesimurex sallasi* (Figure 6.3), the fasciolariid *Aristofusus couei* (Figure 6.14), the busyconids *Busycoarctum coarctatum* (Figure 6.12), *Sinistrofulgur perversum* (Figure 6.10), *Lindafulgur candelabrum* (Figure 6.11), and *Fulguropsis spiratum* (Figure 6.13), the large marginellid *Prunum labiatum* (Figure 6.14), the olivid *Americoliva maya* (Figure 6.18), the cone shell *Gradiconus sennottorum* (Figure 6.20), and the terebrid *Cinguloterebra stegeri* (Figure 6.14). These large gastropods form a classic Yucatan molluscan assemblage that is ubiquitous throughout the subprovince and their presence can be used as an indicator for the subprovincial boundaries.

Besides the widespread Yucatanean endemic taxa, other noteworthy and conspicuous faunal elements that occur throughout the entire subprovince are the widespread Caribbean gastropods, *Melongena melongena* (Figure 6.9) and *Turbinella angulata* (see Petuch and Berschauer, 2021a). These two large species are found in shallow water areas from Veracruz to the Belize-Quintana Roo border, and this stretch of coast is the only area within the Gulf of Mexico region that supports these high-tropical gastropods. Fossil specimens of *Melongena melongena* have been collected in shallow fill pits near Lake

Okeechobee, Florida (Petuch and Berschauer, 2021b; one fossil specimen shown here in Figure 6.9F), demonstrating that this wide-ranging Caribbean species lived on the Floridian Peninsula in the recent past but is now absent from the Floridian Subprovince.

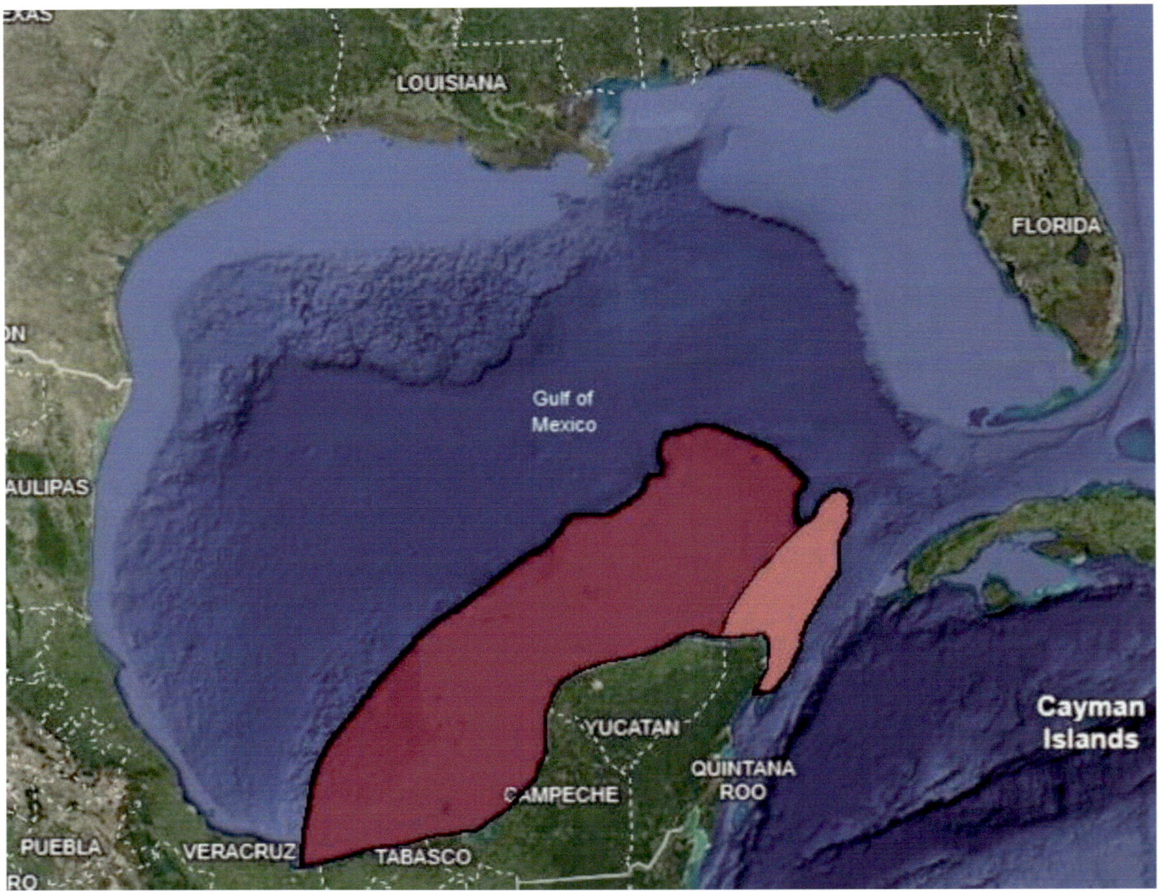

Figure 6.1 Map of the Yucatanean Subprovince (maroon and rose sharing) of the Carolinian Molluscan Province, showing the **Campeche Bank** along the northern and western sides and the **Contoyan Infraprovince** (designated by the rose shading).

Melongena melongena lives in quiet, protected lagoons in close association with Red Mangrove forests and organic-rich mud flats. Populations that inhabit hypersaline pools and enclosed lagoons are often dwarfed and take on a high-spired appearance. These miniature *M. melongena* were given the form name *salinarum*, in recognition of their unusual hypersaline habitat (one individual shown here in Figure 6.9E). The large West Indian Chank Shell, *Turbinella angulata*, occurs on Turtle Grass (*Thalassia*) beds in shallow, often intertidal, depths where it feeds on marine worms. Like *M. melongena*, *T. angulata* is also known from the late Pleistocene fossil beds of southern Florida (Petuch, 1994; 2004), demonstrating that this large turbinellid had a much wider distribution in the very recent past.

Endemic Mollusks of the Coastal Lagoons

Almost the entire coastline of the Yucatan Peninsula is fringed with extensive lagoon systems, with some being almost completely fresh water (oligohaline), some being brackish (hyposaline), and others having evaporite-producing conditions (hypersaline). These water chemistry conditions often vary seasonally, becoming hyposaline during the rainy summer months and hypersaline during the dry winter months. The largest and most important of these coastal lagoons include:

- Laguna de la Machona (in western Tabasco State).
- Rendonda-Tupilco-Mecoacan Lagoon Complex (in eastern Tabasco State).
- Laguna del Carmen (in western Campeche State).
- Petenes-Ria Celestun Lagoon and Wetlands Complex (in northeastern Campeche State).
- El Corchito-Chicxulub-Rosada Lagoon Complex (in western Yucatan State).
- Laguna Ria Lagartos (in eastern Yucatan State).
- Laguna Yalahau-Laguna Holbox Complex (in northern Quintana Roo State).
- Chacmuchuch-Isla Blanca Lagoon Complex (north of Cancun, Quintana Roo State).
- Nichopte Lagoon (at Cancun, Quintana Roo State).

These coastal lagoons all share a common molluscan fauna, most often dominated by a few hardy Yucatanean Subprovince endemic species. The lagoonal endemics are generally euryhaline (able to tolerate a wide range of salinities) and are able to survive the seasonal salinity variations. Some of the hypersaline evaporite lagoons, like the Laguna Rosada of Yucatan State, take on a bright reddish-pink color due to massive growths of unicellular red algae. These red algae support a fauna of tiny crustaceans which are the main food resource for wading birds like Flamingos and Roseate Spoonbills.

Typically, the quiet, sheltered lagoons support extensive meadows of the sea grasses *Thalassia*, *Syringodium*, and *Halodule* and these form the ecological base for an interesting but impoverished grass-associated molluscan fauna with a high level of endemism. Primary among these grass bed endemic mollusks are the modulids *Modulus hunahpu* and *Trochomodulus velai* (both shown in Figure 6.2), the large muricid *Phyllonotus mexicanus* (Figure 6.3), the fasciolariid *Cinctura salasi* (Figure 6.6), the cone shell *Gradiconus maya* (Figure 6.21), and the bubble shell *Bulla morai* (Figure 6.2). This endemic Yucatan molluscan community is an exact analogue of the molluscan assemblage found in the Florida Keys Turtle Grass beds (Floridian Subprovince). There, similar-appearing species such as the modulid *Trochomodulus calusa*, the muricid *Phyllonotus pomum*, the fasciolariid *Cinctura hunteria*, the cone shell *Gradiconus burryae*, and the bubble shell *Bulla frankovichi*, occupy the exact same ecological niches as their Yucatan counterparts and constitute parallel community evolution. The morphological and ecological similarities between these sets of sibling species indicates that they evolved from a single ancestral *Thalassia*-based community, probably during the mid-Pleistocene. The intertidal mud and sand flats associated with these lagoons also support large aggregations of the small melongenid, *Melongena (Rexmela) bispinosa* (Figure 6.7) and its ecophenotypic variant *martiniana* (Figure 6.8). These endemic carnivorous and carrion-feeding gastropods occur together with the larger *Melongena melongena*, the busyconid *Sinistrofulgur perversum*, and the endemic vermivorous cone shells *Jaspidiconus chaac* (Figure 6.19) and *Jaspidiconus ixchel* (Figure 6.23).

Endemic Mollusks of the Campeche Bank

The northern and western sides of the Yucatan Peninsula are bounded by the immense Campeche Bank, a shallow carbonate platform that averages only 20-200 m in depth (Figure 6.1). Spatially, the Campeche Bank is almost the size of the entire Yucatan Peninsula and is the largest purely tropical continental shelf area in the entire Gulf of Mexico. The Bank extends from the Tabasco State border eastward to Cozumel Island, Quintana Roo State and is widest along the area directly off Yucatan State and narrowest along the Yucatan Channel area of eastern Quintana Roo State. The high tropical nature of the Bank is underscored by the presence of many offshore coral reef complexes and archipelagos of coral cays. These geomorphological features are best-developed along the western side of the Campeche Bank and parallel the abrupt drop-off into the deep Bay of Campeche (**Tabasco-Campeche Basin**). Some of the largest and most important of these coralline landforms, collectively referred to as the **Campeche Bank Archipelago**, include (from north to south); Cayo Arenas, the Triangulo Este and Triangulo Oeste Reefs, the Banco Obispo Norte and Banco Obispo Sur Reefs, the Banco Pera Reef, the Banco Nuevo Reefs, and Cayo Arcas. The largest and best-developed coralline structure of the Campeche Bank Archipelago is **Alacranes Reef Atoll**, 125 km due north of Progreso, Yucatan State. Complete with an enclosed central lagoon that is surrounded by a circle of coral cays and barrier reefs, Alacranes Reef represents the only coral atoll found in the Gulf of Mexico. The main islands of the atoll include Desterrada, Desertora, Perez, and the Chica Pajaros, and their flat, sparcely-vegetated surfaces provide shelter for a colony of the Magnificent Frigatebird.

The wide shallow platform of the Campeche Bank was one of the principal shrimping grounds during the 1950s and on into the 1980s, and large numbers of mollusks were saved as bycatch from the shrimp nets. These specimens gave insight into an entirely new type of Gulf fauna and served as the source of the research material for pioneer scientists such as Rehder and Abbott (1951; a new cone, muricid, volute, vase shell, and fasciolariid), and Clench (1953), Emerson (1979), and Garcia and Emerson (1987), who, altogether, described three new endemic volutes. The vast areas of open carbonate sand sea floors that exist between the coral reef complexes and coral cays house a large number of gastropods that are restricted to the Campeche Bank, some of which include the small turritellid *Torculoidella lyonsi* (Figure 6.2), the large banded Tulip Shell *Fasciolaria (Hollisteria) branhamae* (Figure 6.5), the volute *Scaphella butleri* (Figure 6.17), the cancellariids *Ventrilia jonathoni* and *Ventrilia xbalanque* (both in Figure 6.14), the cone shell *Lindaconus therriaulti* (Figure 6.22), and the terebrid *Cinguloterebra stegeri* (Figure 6.14). The Campeche Bank coral reefs and coral rubble sea floors also shelter several interesting endemic species, including the small muricid *Favartia barbarae* (Figure 6.3), and the pisaniid *Engina dicksoni* (Figure 6.14), and the elongated turbinellid *Siphovasum latiriforme* (Figure 6.16). Some of these widespread Campeche Bank species, along with the colorful fasciolariids *Cinctura connori* (Figure 6.4) and *Cinctura tracieae* (Figure 6.24), are common in the offshore ecosystems along Isla Holbox and Isla Catoche, Quintana Roo State, and are prominent organisms within the Contoyan Infraprovince (discussed next).

Endemism in the Contoyan Infraprovince

The extreme eastern edge of the Campeche Bank, along the northeasternmost part of the Yucatan Peninsula from Isla Holbox and Cabo Catoche to Isla Contoy and to the deep water areas on the Catoche Spur (see Figure 0.2), Quintana Roo State, represents an "evolutionary hotspot" where numerous endemic taxa are found within a very small geographical area. This restricted area of special evolution, referred to here as the **Contoyan Infraprovince** (for Isla Contoy), contains a number of remarkable gastropods, some of which are among the largest found within the entire Gulf of Mexico region. Primary among these oversized Contoyan species are the giant volutes *Garciavoluta mcginnorum* (Figure 6.30) and *Caricellopsis contoyensis* (Figure 6.29), which, along with the northern Gulf *Aurinia kieneri*, are the largest volutoideans found in the entire Carolinian Province. These two impressive volutes coexist with several other Contoyan endemics, including their smaller relative, *Scaphella stimpsonorum* (Figure 6.28), the olive shell *Americoliva contoyensis* (Figure 6.18), the marginellids *Prunum holboxiensis* and *Prunum lipei* (Figure 6.19), the fasciolariids *Fasciolaria haleyi* (Figure 6.26), *Fasciolaria ixchel* (Figure 6.31; endemic to the Isla Mujeres area), and *Cinctura (Hollisteria) branhamae morganae* (Figure 6.25; the Contoyan Infraprovince subspecies of the Campeche Bank *C. branhamae*), and the busyconid *Lindafulgur camazotz*.

The Laguna de Yalahau and Isla Holbox area of the Contoyan Infraprovince houses one of the most remarkable gastropods known from the entire Carolinian Province; the giant turbinellid chank shell, *Turbinella wheeleri* (Figure 6.15). This large species is particularly noteworthy in that it was first described as a late Pleistocene (late Tarantian Age) fossil from the Fort Thompson Formation of southern Florida, where it is relatively common in the shallow subsurface beds of the Everglades region (Petuch, 1994, 2004; Petuch and Berschauer, 2021b). There, fossil *Turbinella wheeleri* (shown here in Figure 6.15) is found along with fossils of another common Yucatan endemic species, *Melongena (Rexmela) bispinosa*, and also the Yucatan-Caribbean Sea *Melongena melongena*. This demonstrates that all three species were recently living together in Florida but became extinct there only in the past few tens of thousands of years. The Contoyan molluscan assemblage can now be seen to represent a relictual fauna, having once been distributed over a much wider area than it is now but then dying out everywhere except for the eastern side of the Yucatanean Subprovince. Another distinctive Contoyan endemic is the small cone shell *Cariboconus kirkandersi* (Figure 6.20), which is restricted to Isla Mujeres and Isla Cozumel at the extreme southern end of the Yucatanean Subprovince. This shallow water, rock-dwelling species is the only member of the Caribbean Province genus *Cariboconus* that occurs in the Carolinian Province. Recent studies have shown that *C. kirkandersi* is a member of a species complex of at least eight closely-related siblings that are distributed on isolated islands along the western side of the Caribbean Basin (Petuch and Berschauer, 2021a).

TABLE 7. Important Mollusks that are Endemic to the Yucatanean Subprovince.

Gastropoda

Modulidae

Modulus hunahpu Petuch and Berschauer, 2023
Trochomodulus velai Petuch and Berschauer, 2023

Turritellidae

Torculoidella lyonsi (Garcia, 2006)

Muricidae

Calotrophon hystrix Garcia, 2006
Favartia barbarae E. Vokes, 1994
Phyllonotus mexicanus (Petit de la Saussaye, 1852)
Vokesimurex sallasi (Rehder and Abbott, 1951)

Fasciolariidae

Aristofusus couei (Petit de la Saussaye, 1853)
Cinctura connori Petuch and Berschauer, 2020
Cinctura salasi Petuch and Berschauer, 2023
Cinctura traciae Petuch and Berschauer, 2020
Cinctura (Hollisteria) branhamae (Rehder and Abbott, 1951)
Cinctura (Hollisteria) branhamae morganae Petuch and Berschauer, 2020
Fasciolaria haleyi Berschauer and Petuch, 2023
Fasciolaria ixchel Petuch and Berschauer, 2024

Busyconidae

Busycoarctum coarctatum (Sowerby I, 1825)
Fulguropsis spiratum (Lamarck, 1816)
Lindafulgur camazotz Petuch and Berschauer, 2023
Lindafulgur candelabrum (Lamarck, 1816)
Sinistrofulgur perversum (Linnaeus, 1758)

Melongenidae

Melongena (Rexmela) bispinosa (Philippi, 1844)
Melongena (Rexmela) bispinosa form *martiniana* (Philippi, 1844)

Pisaniidae

Engina dicksoni Petuch, 2013

Turbinellidae

Siphovasum latiriforme Rehder and Abbott, 1951
Turbinella wheeleri Petuch, 1994

Volutidae

Caricellopsis contoyensis Emerson, 1979
Garciavoluta mcginnorum (Garcia and Emerson, 1987)
Scaphella butleri Clench, 1953
Scaphella stimpsonorum Cossignani and Allary, 2019

Cancellariidae

Ventrilia jonathoni Petuch and Berschauer, 2023
Ventrilia xbalanque Petuch and Berschauer, 2023
Marginellidae
Prunum holboxiensis (Menke, 1828)
Prunum labiatum (Kiener, 1841)
Prunum lipei (Clover, 1990)
Olividae
Americoliva contoyensis (Petuch and Sargent, 1986)
Americoliva maya (Petuch and Sargent, 1986)
Conidae
Cariboconus kirkandersi (Petuch, 1987)
Gradiconus maya (Petuch and Sargent, 2011)
Gradiconus sennottorum (Rehder and Abbott, 1951)
Jaspidiconus chaac Petuch, Berschauer, and Poremski, 2017
Jaspidiconus ixchel Petuch, Berschauer, and Poremski, 2017
Lindaconus therriaulti Petuch, 2013
Terebridae
Cinguloterebra stegeri (Abbott, 1954)
Bullidae
Bulla morai Petuch and Berschauer, 2023

Close-up view of two living *Cariboconus kirkandersi* showing the red animals. This small cone species is endemic to Isla Mujeres and Isla Cozumel, Quintana Roo State, Mexico (photo courtesy of Andre Poremski).

ICONOGRAPHY OF YUCATANEAN SUBPROVINCE ENDEMIC SPECIES

Figure 6.2 Endemic Gastropods of the Yucatanean Subprovince.
A, B= *Modulus hunahpu* Petuch and Berschauer, 2023, width 11 mm, washed up on the beach at Celestun, Yucatan State, Mexico; **C=** *Torculoidella lyonsi* (Garcia, 2006), holotype, length 17.8 mm, 40 m depth in the Bay of Campeche, off Campeche, Campeche State, Mexico; **D, E=** *Trochomodulus velai* Petuch and Berschauer, 2023, width 12 mm, washed up on the beach at Celestun, Yucatan State, Mexico; **F, G=** *Bulla morai* Petuch and Berschauer, 2023, holotype, length 18 mm, Ria Celestun, Celestun, Yucatan State, Mexico (originally described as a subspecies of the widespread western Atlantic *Bulla occidentalis*).

Figure 6.3 Endemic Muricid Gastropods of the Yucatanean Subprovince.
A, B= *Phyllonotus mexicanus* (Petit de la Saussaye, 1853), length 70 mm, Laguna de Yalahau, Isla Holbox, Quintana Roo State, Mexico; **C=** *Phyllonotus mexicanus* (Petit de la Saussaye, 1853), length 65 mm, Laguna de Yalahau, Isla Holbox, Quintana Roo State, Mexico; **D, E=** *Favartia barbarae* E. Vokes, 1994, holotype, length 21 mm, Campeche Bank off Progresso, Mexico; **F, G=** *Vokesimurex sallasi* (Rehder and Abbott, 1951), length 44 mm, 50 m depth on the Campeche Bank, south of the Alacranes Reefs, Mexico; **H=** *Calotrophon hystrix* Garcia, 2006, length 29 mm, 50 m depth on the Campeche Bank, Yucatan, Mexico.

Figure 6.4 *Cinctura connori* from the Campeche Bank and Contoyan Infraprovince.
A, B= *Cinctura connori* Petuch and Berschuaer, 2020, length 89 mm; **C, D=** *Cinctura connori* Petuch and Berschauer, 2020, length 90 mm; **E, F=** *Cinctura connori* Petuch and Berschauer, 2020, holotype, length 72.5 mm. All three specimens were trawled from 250 m depth off Cabo Catoche, Quintana Roo State, Mexico. Originally named as a subspecies of *Cinctura lilium* but now considered to be a full species because of the differences in early whorl sculpture.

Figure 6.5 ***Cinctura (Hollisteria) branhamae*** **from the Campeche Bank.**
A, B= *Cinctura (Hollisteria) branhamae* (Rehder and Abbott, 1951), length 131 mm; **C, D=** *Cinctura (Hollisteria) branhamae* (Rehder and Abbott, 1951), length 140 mm. Both specimens trawled from 60 m depth on the Campeche Bank, north of Progreso, Yucatan State, Mexico.

Figure 6.6 *Cinctura salasi* from the northern Yucatan coast.
A, B= *Cinctura salasi* Petuch and Berschauer, 2023, holotype, length 53 mm; **C, D=** *Cinctura salasi* Petuch and Berschauer, 2023, length 77 mm; **E, F=** *Cinctura salasi* Petuch and Berschauer, 2023, length 64 mm. All three specimens were collected as beach specimens along the Celestun Lagoon area, Celestun, Yucatan State, Mexico.

Figure 6.7 *Melongena (Rexmela) bispinosa* from Isla Holbox, Quintana Roo State, Mexico.
A, B= *Melongena (Rexmela) bispinosa* (Philippi, 1844), length 58 mm; **C, D=** *Melongena (Rexmela) bispinosa* (Philippi, 1844), length 55 mm; **E, F=** *Melongena (Rexmela) bispinosa* (Philippi, 1844), banded color form; length 48 mm. All three specimens were collected on mangrove mud flats near Holbox, Isla Holbox, Quintana Roo State, Mexico; **G=** *Melongena (Rexmela) bispinosa* (Philippi, 1844), length 41 mm, fossil specimen collected in a road fill pit at Okeechobee, Florida, from the Coffee Mill Hammock Member of the Fort Thompson Formation, Tarantian Age of the late Pleistocene.

Figure 6.8 ***Melongena (Rexmela) bispinosa*** **form** ***martiniana*** **from Isla Blanca, Quintana Roo State, Mexico.** **A, B**= *Melongena (Rexmela) bispinosa* form *martiniana* (Philippi, 1844), length 45 mm; **C, D**= *Melongena (Rexmela) bispinosa* form *martiniana* (Philippi, 1844), banded form, length 40 mm; **E, F**= *Melongena (Rexmela) bispinosa* form *martiniana* (Philippi, 1844), orange color variant, length 33 mm. All three specimens collected on mud flats on Isla Blanca, Costa Mujeres, Quintana Roo State, Mexico.

Figure 6.9 *Melongena melongena* from the Yucatanean Subprovince.
A, B= *Melongena melongena* (Linnaeus, 1758), length 120 mm, Laguna de Alvarado, Veracruz State, Mexico; **C, D=** *Melongena melongena* form *multispinosa* Dollfuss, 1887, length 103 mm, off Champoton, Campeche State, Mexico; **E=** *Melongena melongena* form *salinarum* Bruggeman-Nannenga and Hummelinck, 1986, length 39 mm, low tide at Zacatal, Laguna de Terminos, Campeche State, Mexico; **F=** *Melongena melongena* (Linnaeus, 1758), length 109 mm, fossil specimen from the latest Pleistocene (late Tarantian Age) Coffee Mill Hammock Member of the Fort Thompson Formation. This specimen was collected in a shallow road fill pit at Okeechobee, Florida and demonstrates that *M. melongena* was living in Florida in the late Pleistocene but is now extinct in that area.

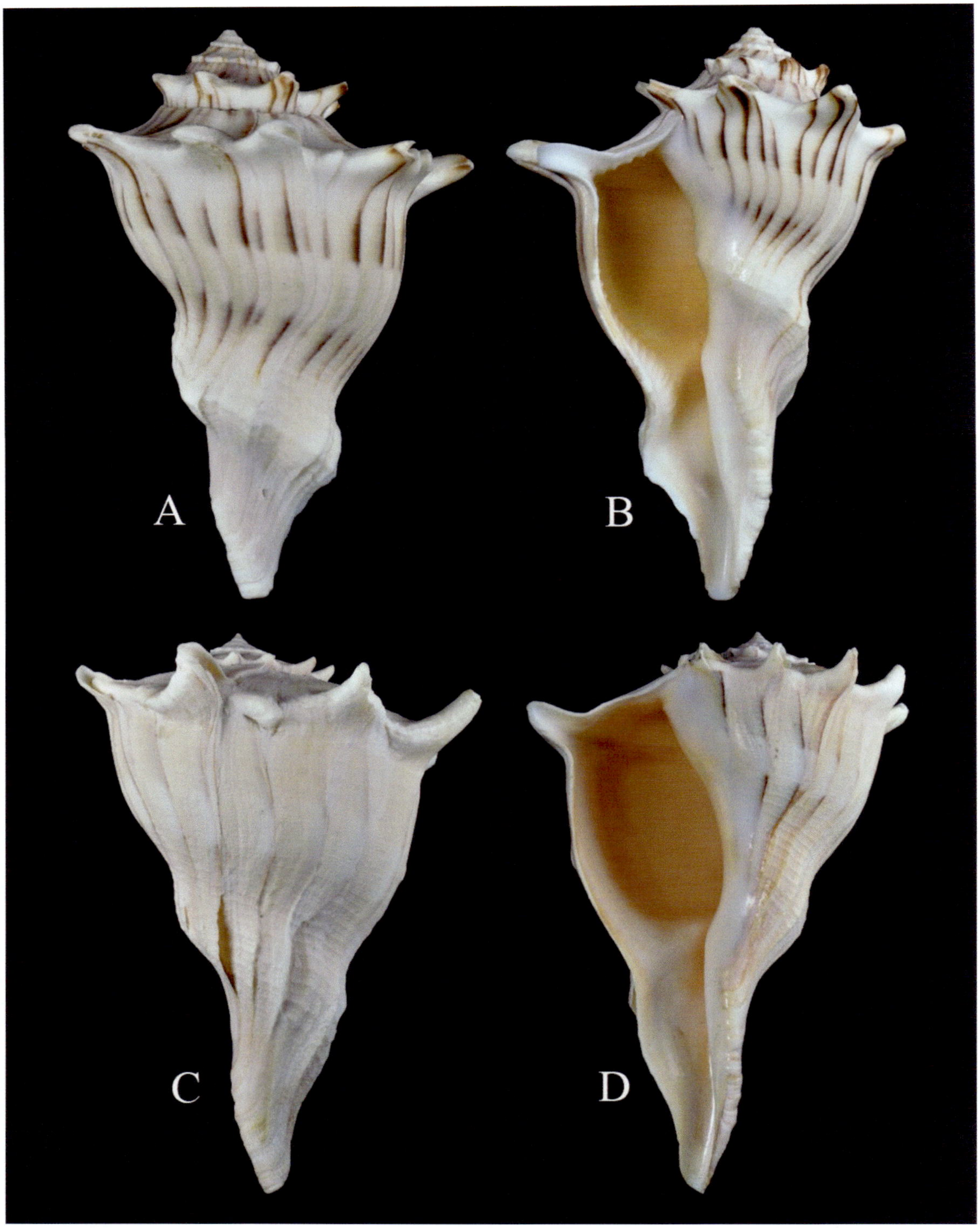

Figure 6.10 ***Sinistrofulgur perversum*** **from the Yucatanean Subprovince.**
A, B= *Sinistrofulgur perversum* (Linnaeus, 1758), length 108 mm; **C, D=** *Sinistrofulgur perversum* (Linnaeus, 1758), length 213 mm. Both specimens were trawled by shrimpers from 20 m depth off Progreso, Yucatan State, Mexico.

Figure 6.11 *Lindafulgur candelabrum* from the Yucatanean Subprovince and southern Texas. A, B= *Lindafulgur candelabrum* (Lamarck, 1816), length 210 mm, 20 m depth east of Veracruz, Veracruz State, Mexico; **C, D=** *Lindafulgur candelabrum* (Lamarck, 1816), specimen with elongated brown flammules on the body whorl and spire, length 177 mm, 20 m depth off Progreso, Yucatan State, Mexico. The interior of the aperture is pure white or very pale yellow, as opposed to the deep orange aperture of the similar *L. camazotz* from the Contoyan Infraprovince (Figure 6.27). This large busyconid also ranges northward to southern Texas.

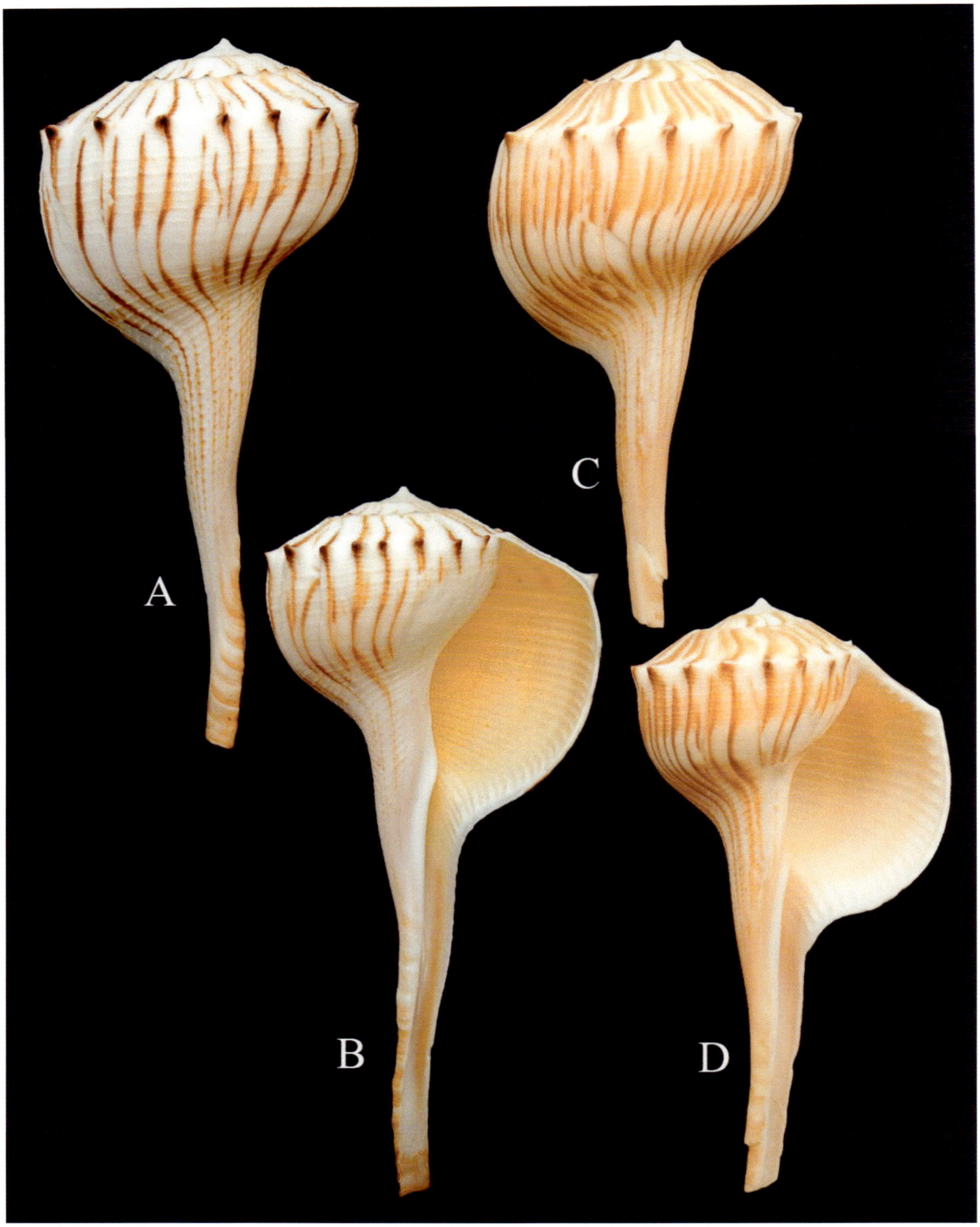

Figure 6.12 ***Busycoarctum coarctatum*** **from the Campeche Bank.**
A, B= *Busycoarctum coarctatum* (Sowerby I, 1825), normal color form, length 138 mm; **C, D=** *Busycoarctum coarctatum* (Sowerby I, 1825), dark color form, length 115 mm. Both specimens were trawled from 30 m depth north of Progreso, Yucatan State, Mexico.

Figure 6.13 *Fulguropsis spiratum* from the Yucatanean Subprovince.
A, B= *Fulguropsis spiratum* (Lamarck, 1816), length 106 mm; **C, D**= *Fulguropsis spiratum* (Lamarck, 1816), golden-orange variant, length 89 mm; **E, F**= *Fulguropsis spiratum* (Lamarck, 1816), length 80 mm. All three specimens were collected on the beach after a storm, at Telchac Puerto, Yucatan State, Mexico.

Figure 6.14 Endemic Taxa from the Campeche Bank and the Bay of Campeche.
A, B= *Engina dicksoni* Petuch, 2013, length 20 mm, 100 m depth off Cabo Catoche, Quintana Roo State, Mexico; **C**= *Aristofusus couei* (Petit de la Saussaye, 1853), apertural view, length 110 mm, 50 m depth south of Alacranes Reef, Campeche Bank, Mexico; **D, E**= *Ventrilia xbalanque* Petuch and Berschauer, 2023, holotype, length 17 mm, 60 m depth off Celestun, Yucatan State, Mexico; **F**= *Cinguloterebra stegeri* (Abbott, 1954), holotype, length 80 mm, 60 m depth off Progreso, Yucatan State, Mexico; **G, H**= *Prunum labiatum* (Kiener, 1841), length 28 mm, 20 m depth off Progreso, Yucatan State, Mexico; **I, J**= *Ventrilia jonathoni* Petuch and Berschauer, 2023, holotype length 12 mm, off Telchac Puerto, Yucatan State, Mexico.

Figure 6.15 Wheeler's Chank Shell from the Contoyan Infraprovince.
A, B= *Turbinella wheeleri* Petuch, 1994, length 280 mm, 20 m depth off Isla Holbox, Quintana Roo State, Mexico; **C, D**= *Turbinella wheeleri* Petuch, 1994, length 345 mm, 20 m depth off Isla Holbox, Quintana Roo State, Mexico (courtesy of Kevin Monsecour); **E**= *Turbinella wheeleri* Petuch, 1994, length 185 mm, fossil specimen from the upper beds of the Fort Thompson Formation, Tarantian Age of late Pleistocene, Palm Beach County, Florida (see Petuch and Berschauer, 2021b).

Figure 6.16 Latiriform Vase Shell from the Yucatanean Subprovince.
A, B= *Siphovasum latiriforme* Rehder and Abbott, 1951, length 44 mm; **C, D**= *Siphovasum latiriforme* Rehder and Abbott, 1951, length 55.7 mm; **E, F**= *Siphovasum latiriforme* Rehder and Abbott, 1951, orange color form, length 34 mm (juvenile shell). All three specimens were found inside coral rubble slabs trawled from 30 m depth off Cabo Catoche, Quintana Roo State, Mexico.

Figure 6.17 *Scaphella butleri* from the Campeche Bank.
A, B= *Scaphella butleri* Clench, 1953, length 114 mm, trawled by shrimpers from 50 m depth off Progreso, Yucatan State, Mexico; **C, D=** *Scaphella butleri* Clench, 1953, length 112 mm, deep reef form, trawled from 60 m depth south of Alacranes Reef, Campeche Bank Archipelago, Yucatan, Mexico.

Figure 6.18 ***Americoliva*** **Species of the Yucatanean Subprovince.**
A, B= *Americoliva maya* (Petuch and Sargent, 1986), pale color form, length 52 mm, found in clean coral sand, 2 m depth off Isla Mujeres, Quintana Roo State, Mexico; **C, D=** *Americoliva maya* (Petuch and Sargent, 1986), length 49 mm, trawled by shrimpers from 20 m depth in the Bay of Campeche, west of Campeche, Campeche State, Mexico; **E, F=** *Americoliva contoyensis* (Petuch, 1988), length 35 mm, 20 m depth off Cabo Catoche, Quintana Roo State, Mexico; **G=** *Americoliva contoyensis* (Petuch, 1988), holotype, length 37 mm, trawled from 35 m depth off Isla Contoy, Quintana Roo State, Mexico.

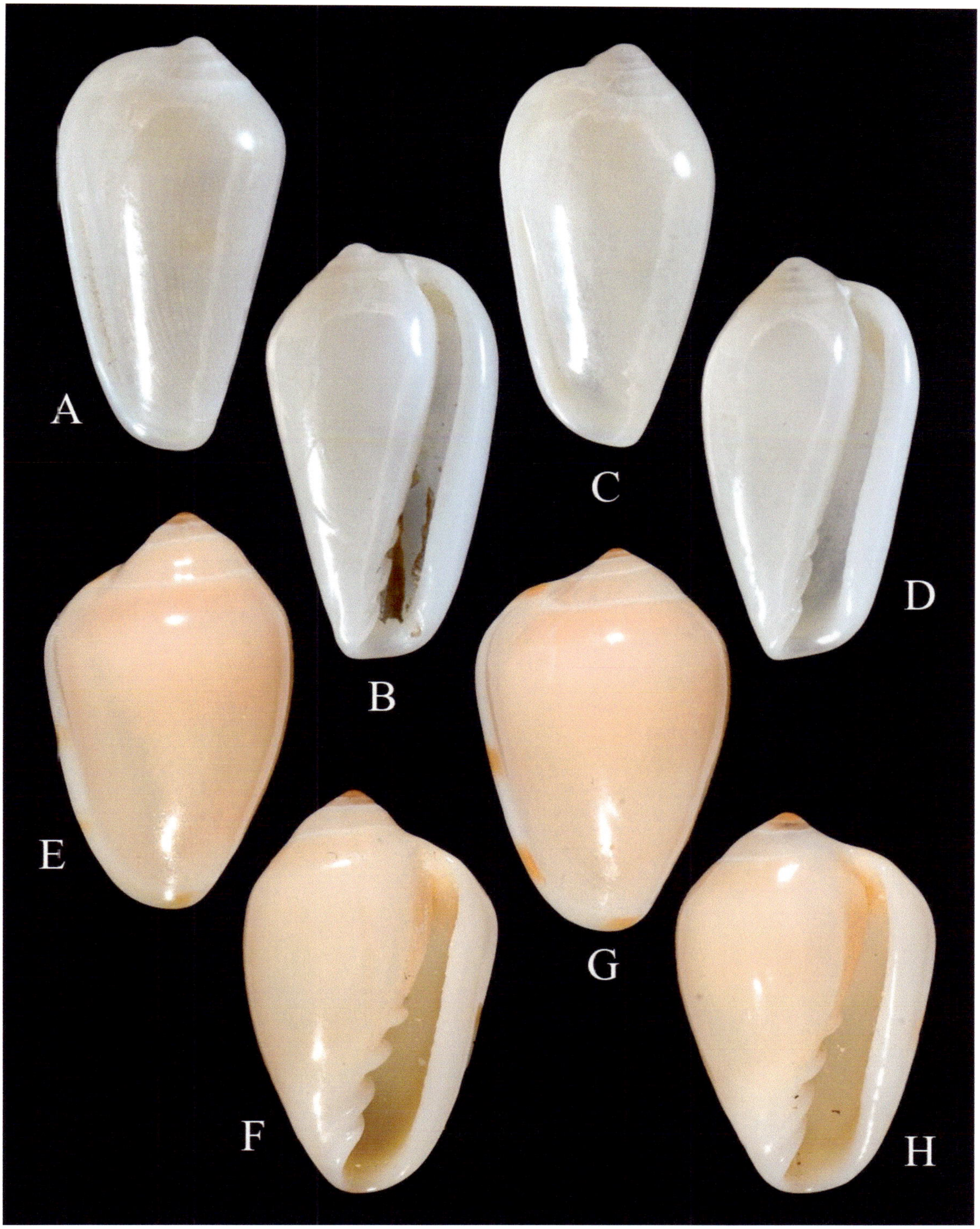

Figure 6.19 Endemic Marginellids from the Yucatanean Subprovince.
A, B= *Prunum lipei* (Clover, 1990), length 9.5 mm, 20 m depth off Isla Contoy, Qintana Roo State, Mexico; **C, D=** *Prunum lipei* (Clover, 1990), length 9.2 mm, 20 m depth off Isla Contoy, Quintana Roo State, Mexico; **E, F=** *Prunum holboxiensis* (Menke, 1828), length 11.4 mm, 2 m depth in Laguna Yalahau, Isla Holbox, Quintana Roo State, Mexico; **G, H=** *Prunum holboxiensis* (Menke, 1828), length 12.0 mm, 2 m depth in Laguna Yalahau, Isla Holbox, Quintana Roo State, Mexico.

Figure 6.20 Cone Shells of the Yucatanean Subprovince.
A, B= *Cariboconus kirkandersi* (Petuch, 1987), length 15 mm; **C, D=** *Cariboconus kirkandersi* (Petuch, 1987), length 16 mm; **E=** *Cariboconus kirkandersi* (Petuch, 1987), length 15 mm (all three specimens of *Cariboconus kirkandersi* were collected in 1-2 m depths on the northern tip of Cozumel Island, Quintana Roo State, Mexico; photos of specimens courtesy of Andre Poremski); **F, G=** *Gradiconus sennottorum* (Rehder and Abbott, 1951), length 35 mm, 50 m depth south of the Alacranes Reef, Campeche Bank, Mexico.

Figure 6.21 *Gradiconus maya* from the Yucatanean Subprovince.
A, B= *Gradiconus maya* Petuch and Sargent, 2011, length 46 mm; **C, D=** *Gradiconus maya* Petuch and Sargent, 2011, length 28 mm; both specimens were collected near Turtle Grass beds within the Laguna de Yalahau, Isla Holbox, Quintana Roo State, Mexico; **E, F=** *Gradiconus maya* Petuch and Sargent, 2011, length 34.3 mm, collected near Celestun Lagoon, Celestun, Yucatan State, Mexico.

Figure 6.22 *Lindaconus therriaulti* from the Campeche Bank.
A, B= *Lindaconus therriaulti* Petuch, 2013, length 62 mm; **C, D=** *Lindaconus therriaulti* Petuch, 2013, length 63 mm; **E, F=** *Lindaconus therriaulti* Petuch, 2013, length 41.6 mm. All three specimens were trawled by shrimpers from 20 m depth on the Campeche Bank off Progreso, Yucatan State, Mexico. May also come in a pure white color or with orange spots.

Figure 6.23 Endemic *Jaspidiconus* Species from the Yucatanean Subprovince.
A, B= *Jaspidiconus ixchel* Petuch, Berschauer, and Poremski, 2017, holotype, length 15 mm, from off Isla Mujeres, Quintana Roo State, Mexico; **C, D=** *Jaspidiconus ixchel* Petuch, Berschauer, and Poremski, 2017, corded form, length 16 mm, from off Isla Mujeres, Quintana Roo State, Mexico; **E, F=** *Jaspidiconus chaac* Petuch, Berschauer, and Poremski, 2017, holotype, length 15 mm, 2.5 m depth off Playa del Carmen, Quintana Roo State, Mexico; **G, H=** *Jaspidiconus chaac* Petuch, Berschauer, and Poremski, 2017, pale color form, length 15 mm, 2.5 m depth off Playa del Carmen, Quintana Roo State, Mexico.

Figure 6.24 *Cinctura traciae* from the Campeche Bank and Contoyan Infraprovince.
A, B= *Cinctura traciae* Petuch and Berschauer, 2020, length 72 mm; **C, D=** *Cinctura traciae* Petuch and Berschauer, 2020, holotype, length 72.3 mm; **E, F=** *Cinctura traciae* Petuch and Berschauer, 2020, length 75.0 mm. All three specimens were trawled from 250 m depth off Cabo Catoche, Quintana Roo State, Mexico. Originally considered a subspecies of *Cinctura tortugana* but now given full species status because of the great differences in early whorl sculpture.

Figure 6.25 *Cinctura (Hollisteria) branhamae morganae* from the Contoyan Infraprovince.
A, B= *Cinctura (Hollisteria) branhamae morganae* Petuch and Berschauer, 2020, length 113 mm; **C, D=** *Cinctura (Hollisteria) branhamae morganae* Petuch and Berschauer, 2020, length 89.2 mm; **E, F=** *Cinctura (Hollisteria) branhamae morganae* Petuch and Berschauer, 2020, holotype, length 71.2 mm. All three specimens were trawled from 200 m depth off Cabo Catoche, Quintana Roo State, Mexico.

Figure 6.26 *Fasciolaria haleyi* from the Contoyan Infraprovince.
A, B= *Fasciolaria haleyi* Berschauer and Petuch, 2023, holotype, length 145.6 mm; **C, D=** *Fasciolaria haleyi* Berschauer and Petuch, 2023, length 148 mm; **E=** *Fasciolaria haleyi* Berschauer and Petuch, 2023, length 147.9 mm; **F=** *Fasciolaria haleyi* Berschauer and Petuch, 2023, length 149 mm. All specimens were trawled from 20 m depth off Cabo Catoche, Quintana Roo State, Mexico.

Figure 6.27 *Lindafulgur camazotz* from the Contoyan Infraprovince.
A, B= *Lindafulgur camazotz* Petuch and Berschauer, 2023, holotype, length 208 mm; **C, D=** *Lindafulgur camazotz* Petuch and Berschauer, 2023, length 103 mm. Both were trawled from 200 m depth off Cabo Catoche, Quintana Roo State, Mexico. *Lindafulgur camazotz* is similar to the Suwannean Subprovince *L. lyonsi* (Petuch, 1987) (Figure 4.24, Chapter 4), but differs in having a thicker, heavier, and more colorful shell with more numerous and much stronger spiral cords on the body whorl and spire whorls.

Figure 6.28 *Scaphella stimpsonorum* from the Contoyan Infraprovince.
A, B= *Scaphella stimpsonorum* Cossignani and Allary, 2019, length 123 mm; **C, D**= *Scaphella stimpsonorum* Cossignani and Allary, 2019, length 72 mm (localty data given as "Alacranes Reef, Campeche Bank" in Petuch and Berschauer, 2021a; however, that data was later found to be erroneous and that the illustrated specimen actually came from off Cabo Catoche). Both were trawled from 200 m depth off Cabo Catoche, Quintana Roo State, Mexico. A low-spired, inflated variant, closely resembling the Dry Tortugas *S. elizabethae*, has been trawled in deep water off Isla Mujeres, Quintana Roo State.

Figure 6.29 *Caricellopsis contoyensis* from the Contoyan Infraprovince.
A, B= *Caricellopsis contoyensis* (Emerson, 1979), length 164 mm; **C, D=** *Caricellopsis contoyensis* (Emerson, 1979), length 187.5 mm. Both specimens were trawled from 200 m depth off Cabo Catoche, Quintana Roo State, Mexico.

Figure 6.30 *Garciavoluta mcginnorum* from the Contoyan Infraprovince.
A, B= *Garciavoluta mcginnorum* (Garcia and Emerson, 1987), length 213 mm; **C, D**= *Garciavoluta mcginnorum* (Garcia and Emerson, 1987), holotype, length 158 mm. Both specimens were trawled from 300 m depth off Contoy Light, Isla Contoy, Quintana Roo State, Mexico. This is the type species of the monotypic genus *Garciavoluta*.

Figure 6.31 ***Fasciolaria ixchel*** **from Isla Mujeres.**
A, B= *Fasciolaria ixchel* Petuch and Berschauer, 2024, holotype, length 185.0 mm; **C, D=** *Fasciolaria ixchel* Petuch and Berschauer, 2024, length 175.0 mm. Both specimens were trawled by fishermen from 25 m depth off Isla Mujeres, Quintana Roo State, Yucatan Peninsula, Mexico. Although superficially similar to the widespread *Fasciolaria tulipa* (Figure 1.12 in this book), this large, newly-described species differs in having noticeably stepped spire whorls and early whorls that exhibit heavy ornamentation comprised of large rounded knobs and strong spiral cords. It is part of a regional species complex that also includes *Fasciolaria bullisi* (Figure 2.8), *Fasciolaria bittneri* (Figure 3.27), and *Fasciolaria haleyi* (Figure 6.26),

Figure 6.32 Protoconchs and early whorls of the Gulf of Mexico *Fasciolaria* species complex.
A= *Fasciolaria tulipa*, from Turtle Key, Ten Thousand Islands; **B**= *Fasciolaria bittneri*, from the Dry Tortugas; **C**= *Fasciolaria bullisi*, from eastern Yucatan; **D**= *Fasciolaria haleyi*, from off Contoy Light, Yucatan; **E**= *Fasciolaria ixchel*, from Isla Mujeres, Yucatan. These Gulf endemic sibling species all have very different protoconch structures and early whorl sculpture patterns and form a remarkable evolutionary radiation that has only recently been recognized (see Petuch and Berschauer, 2024).

Figure 6.33 Protoconchs and early whorls of Yucatanean Subprovince *Cinctura* species.
A= *Cinctura connori*, from 250 m depth off Cabo Catoche, Quintana Roo State; **B**= *Cinctura traciae*, from 250 m depth off Cabo Catoche, Quintana Roo State; **C**= *Cinctura (Hollisteria) branhamae,* from 60 m depth off Progreso, Yucatan State; D= *Cinctura (Hollisteria) branhamae morganae*, from 200 m depth off Cabo Catoche, Quintana Roo State.

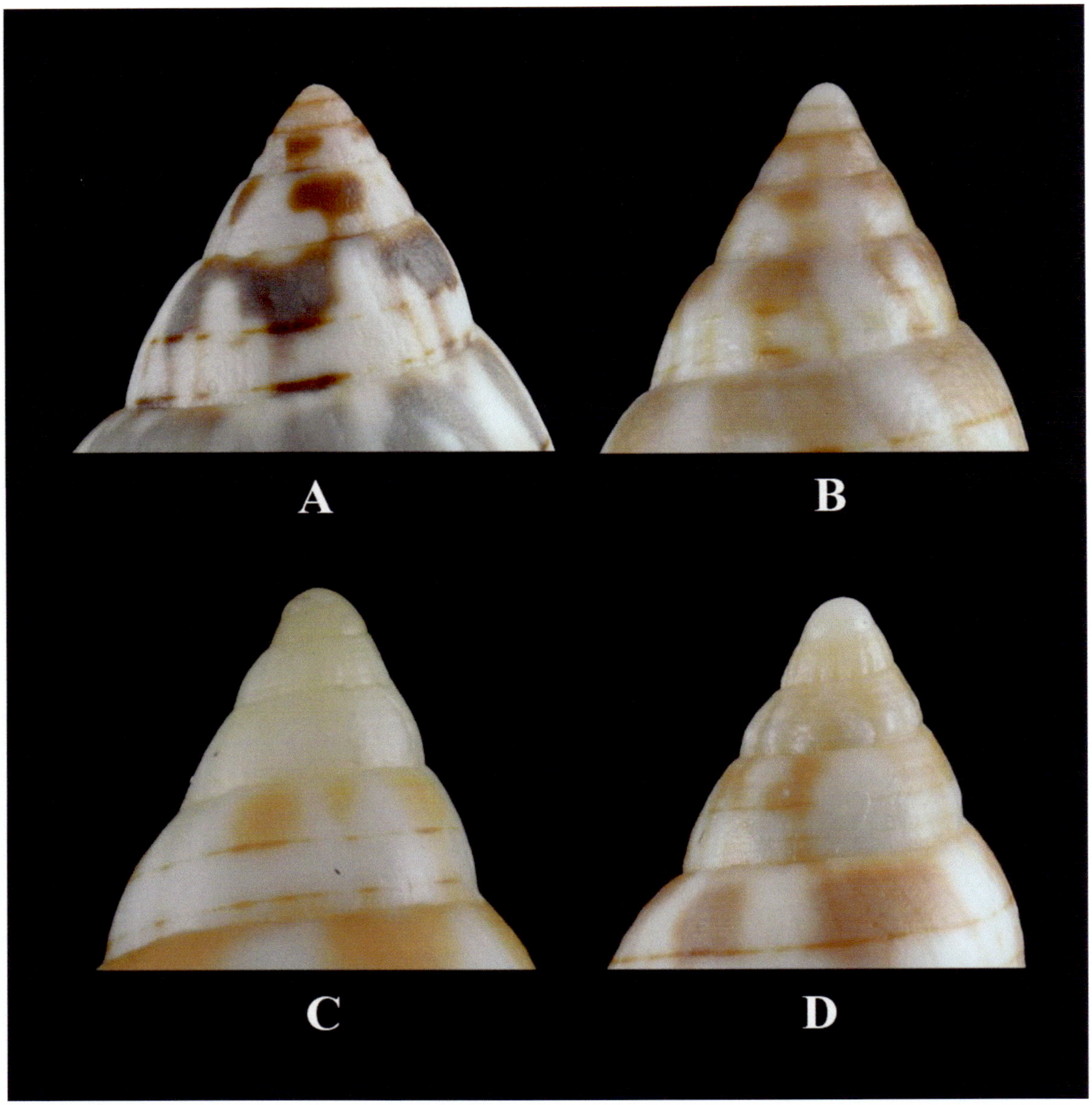

Figure 6.34 Protoconchs and early whorls of *Cinctura* species from the Gulf of Mexico.
A= *Cinctura hunteria*, from Turtle Key, Ten Thousand Islands, Florida; **B=** *Cinctura hunteria apalachee*, from St. Andrew Bay, Bay County, Florida; **C=** *Cinctura tortugana*, from 100 m depth off the Dry Tortugas, Florida; **D=** *Cinctura lilium*, from Port Aransas, Texas. These Florida and Texas Banded Tulip Shells are shown here for comparison with the Yucatanean Subprovince *Cinctura* fauna.

Systematic Appendix

In their 2011 book, *Rare and Unusual Shells of the Florida Keys and Adjacent Areas* (MdM Publishers), Petuch and Sargent described seven important new species and a new volutid subgenus from the Floridian and Suwannean Subprovinces. These biogeographically-important taxa are illustrated in the figures shown here in Chapters 3 and 4. Since the older 2011 book is now out of print and difficult to acquire, we have decided to reprint the original descriptions in an appendix to this new book in order to facilitate the research efforts of future malacologists. Two other new species were described in the 2011 book but are not included here as they are from areas outside of the scope of this book. These include: *Latirus (Polygona) williamlyonsi* (Family Fasciolariidae) collected off Port Pierce, Florida and endemic to the Georgian Subprovince of the Carolinian Province; and *Scaphella (Caricellopsis) matchetti* (Family Volutidae), supposedly from 200 m depth west of Naples, Florida. That locality data, received from a local shell dealer, is now known to be in error and this small volute is actually from deep water off northern Colombia, South America (from the Venezuelan Subprovince of the Caribbean Province).

The following descriptions and discussions are taken from Petuch and Sargent (2011: 143-151). The holotypes of the new taxa are deposited in the Division of Mollusks, Department of Invertebrates, National Museum of Natural History, Smithsonian Institution, Washington, D.C. and bear USNM catalog numbers. The species and subspecies re-described here include:

Pusula juyingae (illustrated in Chapter 3, Figure 3.6 of this book)
Vokesimurex morrisoni (illustrated in Chapter 1, Figure 1.4 of this book)
Phyllonotus whymani (illustrated in Chapter 4, Figure 4.23 of this book)
Favartia goldbergi (illustrated in Chapter 3, Figure 3.9 of this book)
Gemophos tinctus pacei (illustrated in Chapter 4, Figure 4.27 of this book)
Scaphella junonia elizabethae (illustrated in Chapter 3, Figure 3.14 of this book)
Tritonoharpa janowskyi (illustrated in Chapter 3, Figure 3.26 of this book)
The new subgenus ***Caricellopsis*** is also re-described.

Classification
Class Gastropoda
Subclass Orthogastropoda
Superorder Caenogastropoda
Order Sorbeoconcha
Suborder Hypsogastropoda
Infraorder Littorinimorpha
Superfamily Velutoidea
Family Triviidae
Genus *Pusula* Jousseaume, 1884

Pusula juyingae **Petuch and Sargent, 2011**

Description: Shell large for genus, broad, flattened, almost circular in outline; margins thickened, flaring; dorsum ornamented with 16 large, beaded cords that radiate from dorsal furrow; some dorsal cords bifurcate, especially along margins; dorsal furrow proportionally wide and deep, bordered by 10 large, prominent knoblike beads; aperture narrow, straight, widenening and curving slightly at anterior end; fossula large, covering one-third of aperture, slightly concave; base ornamented with 16 large, evenly-spaced smooth ribs; dorsal shell color pale tan with 2 large, prominent brown patches, one on either side of central region of dorsal furrow; large beads bordering furrow white; dorsal radiating cords marked with small, evenly-spaced brown dots; posterior end of dorsal furrow marked with small brown patch; color of shell base pale tan with lighter whitish ribs.
Holotype: length 16.7 mm, USNM 1152540.
Type Locality: Dredged from 120 m depth off Palm Beach Island, Palm Beach County, Florida.
Etymology: Named for Juying Janowsky of Wellington, Florida.
Discussion: The new deep reef species is the largest member of the family Triviidae found in Florida and the northern Caribbean. Only *Pusula labiosa* (Gaskoin, 1836) from the southern Caribbean is larger (illustrated in Petuch, 1987: plate 22, figures 15 and 16). This large species has recently, incorrectly, been referred to the taxon *Pusula costispunctata* (Sowerby II, 1870) a similar-appearing species from the Eastern Pacific Panamic Province (the type locality is here confined to Gobernadora Island, Panama; Schilder's *circumdata* from Ecuador is the same species as *costispunctata*). Online shell identification sites, such as "Let's Talk Shells" and "Jaxshells", have misidentified the Floridian Subprovince *P. juyingae* as *J. costispunctata* and failed to mention Panamic specimens of that taxon. This mistake has been adopted by the online website "World Register of Marine Species (WoRMS)" and an incorrect type locality was given for *J. costispunctata* (erroneously as "Key West", without examining Sowerby's original specimen and locality data).

Infraorder Neogastropoda
Superfamily Muricoidea
Family Muricidae
Subfamily Muricinae
Genus *Vokesimurex* Petuch, 1987

Vokesimurex morrisoni **Petuch and Sargent, 2011**

Description: Shell of average size for genus, with globose, inflated body and elongated narrow siphonal canal; body whorl and spire whorls with 3 thick, rounded varices per whorl; shoulder rounded; intravarical areas with 3 thin longitudinal ribs; body whorl with 18 evenly-spaced spiral cords, with 6 cords being largest and with subsidiary smaller

cords in-between; cords on subsutural area thinner and more closely-packed; spiral cords intersect varices and longitudinal ribs, producing beaded, knobby pattern; siphonal canal sculptured with 16 spiral cords, with 2 cords being larger than others; small spines sometimes present on varices, usually one at shoulder, and one or two at anterior end; some specimens completely spineless; aperture proportionally large, oval in shape; interior of aperture with 8-10 large knoblike teeth along inner edge of lip; peristome erect, thin and bladelike; shell color varying from bright orange to orange-pink (as on holotype) to bright pink; interior of aperture white.
Holotype: Length 39.5 mm, USNM 1152539.
Type Locality: Dredged from 100 m depth off Singer Island, Palm Beach County, Florida.
Etymology: Named for Robert Morrison of Sarasota, Florida.
Discussion: This colorful new murex has, for over 30 years, been referred to by researchers and collectors as "*Murex bellegladeensis* E. Vokes, 1963" (now placed in the genus *Vokesimurex* Petuch, 1994). The real *Vokesimurex bellegladeensis*, however, is a mid-Pleistocene fossil shell that is ancestral to *V. morrisoni*, but differs in being a much more coarsely-sculptured shell with proportionally much thicker and wider varices and intravarical longitudinal ribs, in having a thicker and stubbier siphonal canal, and in having a proportionally much smaller aperture (specimens illustrated in Petuch, 1994; Petuch and Myers, 2014; Petuch and Berschauer, 2021b).

Genus *Phyllonotus* Swainson, 1833

***Phyllonotus whymani* Petuch and Sargent, 2011**

Description: Shell small for genus, rotund, inflated, with rounded shoulder and sloping subsutural area; spire proportionally low, subpyramidal; siphonal canal proportionally long, broad, recurved posteriorly; body whorl and spire whorls with 3 large, thin varices per whorl; varices recurved laterally; intravarical areas with 2 thick, prominent longitudinal ribs; body whorl encircled with 6 thin spiral cords, with 10 finer, smaller cords being present in-between larger cords; spiral cords strongly ornamented with large elongated beads, giving shell rough texture; entire shell overlaid with sculpture of extremely fine, almost microscopic, spiral threads giving shell frosted appearance; aperture proportionally small for genus, almost round; edge of lip with 14 large pointed teeth along peristome; parietal shield proportionally small, thin, adherent, open and flattened along entire posterior part; columellar part of parietal shield with 6-8 low, rounded teeth; siphonal canal with 8 large spiral threads and ornamented with up to 3 small, flattened spines; shell color bright pink-orange overlaid with 3 thin bands of darker orange-brown, one along shoulder, one around mid-body, and one around anterior end; varices white or pale orange-white with 3 orange-tan bands; interior of aperture pale orange-white; edge of lip white; large brown patch present on anal canal.

Holotype: Length 43.4 mm, USNM 1152533; paratype length 43.8 mm, USNM 1152534.
Type Locality: From a coralline algal substrate in 35 m depth, north of Loggerhead Key, Dry Tortugas, western Florida Keys (collected in commercial lobster traps).
Etymology: Named for Jeff Whyman of Boynton Beach, Florida, who collected the type material from lobster traps in the Florida Keys.
Discussion: The small bright orange and pink *Phyllonotus whymani* is closely related to the common shallow water *P. pomum* (Gmelin, 1791) but differs in being a consistently smaller, less inflated shell with a lower spire, in having a more-reduced spiral sculpture; resulting in less-ornamented and more-rounded varices, in having a proportionally smaller and less-flaring aperture, and in having a smaller, less-developed parietal shield that is adherent and lacks the curled, erect peristomal edge that is so characteristic of *P. pomum*.

Subfamily Muricopsinae
Genus *Favartia* Jousseaume, 1880

Favartia goldbergi **Petuch and Sargent, 2011**

Description: Shell very small for genus, inflated, globose, with distinctly rounded shoulder; spire proportionally low, subpyramidal; body whorl with 4-5 wide, prominent rounded varices; siphonal canal proportionally short, stubby, recurved posteriorward; body whorl sculptured with 6 large, very thick, rounded, knobby spiral cords; intervarical areas relatively smooth, ornamented with fine longitudinal striae and tiny frills; siphonal canal with 2 wide spiral cords; entire shell covered with chalky, smooth intriticalx; aperture proportionally large, almost round; protoconch proportionally large, domelike, composed of one and one-half whorls; shell color pale cream-white; interior of aperture white.
Holotype: Length 10.5 mm, USNM 1152532.
Type Locality: Dredged from 100 m depth off Singer Island, Palm Beach County, Florida.
Etymology: Named for Richard Goldberg of New York, New York.
Discussion: The deep water *Favartia goldbergi* is similar in shape and general form to the common shallow water *F. cellulosa* (Conrad, 1846) but differs in being a much smaller, more inflated and fragile shell, in having a distinctly lower spire, in having smoother, unfrilled varices, in having wider and knobbier spiral cords around the body whorl, and in having a smooth, thick intriticalx.

Superfamily Buccinoidea
Family Pisaniidae
Genus *Gemophos* Olsson and Harbison, 1953

***Gemophos tinctus pacei* Petuch and Sargent, 2011**

Description: Shell of average size for genus, fusiform and biconic, with rounded, sloping shoulder and steeply-sloping subsutural area; spire elevated, protracted, with sloping sides; siphonal canal obsolete, open, ornamented with 6 small, thin cords; body whorl ornamented with 12-13 large, thick, highly-raised spiral cords, with smaller, thinner spiral cord between each pair of larger cords; body whorl and spire whorls also ornamented with 12-14 low, undulating longitudinal ribs; intersection of large spiral rib and longitudinal rib producing large elongated bead; beads prominent, arranged in linear, cancellate pattern; aperture large, approximately one-half shell length, oval in shape; posterior end of aperture and anal canal area with 2 large teeth; inner edge of lip with 12-14 large, elongated ribs that extend into aperture; columella with small, smooth shield and single large fold at anterior end; shell color reddish-brown with scattered prominent, large, bright blue patches; elongated beads on body whorl and spire pale cream-white; interior of aperture white.
Holotype: Length 25.4 mm, USNM 1152538.
Type Locality: On oyster bank at low tide, Goodland Bay, Ten Thousand Islands, Collier County, Florida.
Etymology: Named for Robert Pace of Miami, Florida, who collected the holotype in Goodland Bay.
Discussion: The Ten Thousand Islands *Gemophos tinctus pacei* is most similar to, and is often confused with, the widespread western Atlantic nominate subspecies, *Gemophos tinctus tinctus* (Conrad, 1846). The new southwestern Florida endemic differs from the widespread *G. tinctus tinctus* in having more numerous longitudinal ribs on the body whorl, in having more numerous and stronger beads on the body whorl and spire whorls, and in lacking the wide, beaded parietal shield on the columellar area. With its heavily-beaded, cancellate appearance and fusiform shell outline, the new Ten Thousand Islands species superficially resembles the confamilial *Solenosteira cancellarius* (Conrad, 1846) from the Gulf of Mexico. Further study may show that *G. tinctus pacei* is a full sibling species, distinct from the widespread western Atlantic *tinctus*. (Authors' Note: Petuch and Berschauer, 2022, illustrate several specimens of *G. tinctus pacei* and discuss its ecology; living specimens are shown *in situ* on the *Petaloconchus nigricans* vermetoherms ("reefs") in the Ten Thousand Islands.)

Superfamily Volutoidea
Family Volutidae
Genus *Scaphella* Swainson, 1832

***Scaphella junonia elizabethae* Petuch and Sargent, 2011**
(Authors' Note: Now considered to be a full distinct species, part of the *Scaphella junonia* sibling species complex.)

Description: Shell of similar size to nominate subspecies, thin and inflated, elongate-fusiform, smooth and glossy; shoulder well-developed, slightly angled; spire elevated, subpyramidal; aperture proportionally wide; first 3 postnuclear whorls ornamented with 20-24 faint, low longitudinal ribs; slight constriction present on ribs below suture; ribs become obsolete on later whorls; columella with 4 strong plicae; shell color white to pale pinkish-cream, overlaid with 7-13 rows of widely-spaced light brown, oval-to-round, proportionally-small spots; edge of shoulder with single row of large rectangular checkers; columella and interior of aperture pale yellow-orange.
Holotype: Length 106.8 mm, USNM 1152529.
Type Locality: On open sand near coral, 17 m depth north of Loggerhead Key, Dry Tortugas, western Florida Keys.
Etymology: Named for Elizabeth Pereira-Cardenas of Miami, Florida.
Discussion: The new taxon is proposed as a subspecies of the widespread deep water Carolinian Province *Scaphella junonia* (Lamarck, 1804). The shallow water, reef-associated subspecies *elizabethae* differs consistently from typical *junonia* in being a more inflated shell with a better-developed and more-angled shoulder, in having less-sculptured and smoother early whorls, and in having a less-colorful shell, with fewer, smaller, and more widely-spaced brown spots. Both *Scaphella junonia elizabethae* and the Yucatan endemic *Scaphella junonia butleri* Clench, 1953 share this same type of color pattern composed of small, round spots and the two subspecies superficially resemble each other. The Dry Tortugas *elizabethae*, however, differs from the Yucatan *butleri* in consistently being a smaller shell with a broader and better-developed shoulder, in having a lower spire, and in having smoother early whorls that lack any strong ribbed or reticulated sculpture pattern.

Genus *Scaphella* Swainson, 1832
Subgenus *Caricellopsis* Petuch and Sargent, 2011

***Caricellopsis* Petuch and Sargent, 2011 (described here as a subgenus of *Scaphella*)**

Diagnosis: The taxon is proposed as a subgenus of *Scaphella* Swainson, 1832, including species with small, highly-inflated, bubble-like shells with proportionally very large, dome-shaped calcarellas and low spires. Members of the new subgenus are highly-ornamented with a cancellate sculpture pattern composed of intersecting coarse longitudinal and spiral threads, producing a rough woven texture like that seen on many cancellariid gastropods. This cancellate sculpture is always present at some time during the development of the shell, with some species having the sculpture only on the early whorls while others retain the sculpture into adulthood. The columella has only 2 or 3 thin, poorly-developed plicae, with the anteriormost blending directly into the anterior end of the shell; color pattern usually composed of scattered light brown or yellow checker-shaped spots or large amorphous brown flammules and spiral bands.
Type Species: *Scaphella (Caricellopsis) matchetti* Petuch and Sargent, 2011 (originally described as a Florida Gulf species, but later emended when the correct locality data was

received. This small species is now known to inhabit the deep water areas off northern Colombia, particularly within the Golfo de Morrosquillo, accordingly it should be removed from the list of Gulf of Mexico volutes).

Other Species in *Caricellopsis*: *Caricellopsis evelina* Bayer, 1971, deep water off Panama and northern Colombia, from the San Blas Islands and Golfo de Uraba to Cabo la Vela; *Caricellopsis contoyensis* Emmerson and Old, 1979, deep water off eastern Yucatan, Mexico; *Caricellopsis) luizcoutoi* J. Coltro, 1998, deep water off the Bahamas and Turks and Caicos; *Caricellopsis neptunia* (Clench and Aguayo, 1940, deep water off Jamaica.

Etymology: "Resembling a *Caricella*". Named for the members of this group's resemblance to the extinct Paleogene scaphelline volute genus *Caricella*.

Discussion: The members of this subgenus closely resemble members of the extinct genus *Caricella*, particularly in having large, domelike calcarellas and in having a coarse cancellate sculpture on the early whorls. In this last shell character, the *Caricellopsis* species also resemble members of the caricelliform genera *Reticulacella* and *Atraktus*.

Superfamily Cancellarioidea
Family Cancellariidae
Subfamily Plesiotritoninae
Genus *Tritonoharpa* Dall, 1908

Tritonoharpa janowskyi Petuch and Sargent, 2011

Description: Shell small for genus, stocky, proportionally wide, broad across shoulder and mid-body; spire elevated but proportionally low for genus; shell dorsoventrally flattened, with 2 large, thick varices per whorl; varices with distinctive wing-like projection on posterior end; body whorl and spire with strong cancellate sculpture; areas between varices with 30 thin longitudinal riblets; body whorl encircled with 20 thin spiral cords; intersections of spiral corda and longitudinal riblets producing small rounded bead, giving shell rough-textured appearance; aperture proportionally wide, oval in shape; parietal shield free-standing, developed into collar; siphonal canal short, wide, open; shell color white, overlaid with large rectangular brown spot at midline; aperture and parietal shield white; protoconch proportionally large, rounded, dark purple in color.

Holotype: Length 15.8 mm, USNM 1152535.

Type Locality: Dredged from 120 m depth off Palm Beach Island, Palm Beach County, Florida.

Etymology: Named for Robert Janowsky of Wellington, Florida.

Discussion: This small deep water cancellariid is most similar to the common shallow water, reef-dwelling *Tritonoharpa lanceolata* (Menke, 1828), but differs in being a less-elongated, much stockier, wider shell that is broader across the mid-body, in being dorsoventrally flattened; and in having better-developed varices with prominent wing-like extensions on the posterior ends.

Sthenorhytis pernobilis (Fischer and Bernardi, 1857), length 45 mm; from 300 m depth off the Dry Tortugas, Florida Keys. Although widespread over the Caribbean Molluscan Province and the Gulf of Mexico, this magnificent species is rarely collected, primarily due to its deep water habitat. Within the Gulf of Mexico, it is known to be a component of the Sigsbee Escarpment ecosystems.

Literature Cited

Abbott, R.T. 1974. *American Seashells* (2nd Edition). Van Nostrand Reinhold Publishers, New York. 663 pp.

Abbott, R.T and G. Warmke. 1961. *Caribbean Seashells*. Livingston Publishing Company, Narberth, Pennsylvania. 348 pp.

Berschauer, D.P. 2022. The True Identity of *Gradiconus anabathrum* (Crosse, 1865). *Xenophora Taxonomy* 35:3-8.

Bright, T.J. and L.H. Pequegnat. 1974. *Biota of the West Flower Garden Bank*. Gulf Publishing Company, Houston, Texas. 285 pp.

Clarke, A.H. 1989. New Mollusks from undersea oil seeps off Louisiana. *Malacology Data Net* 2(5-6):122-134.

Dekkers, A. and L. Ros. 2022. On the invasion of *Naria turdus* (Gastropoda, Cypraeidae) in the Dutch Caribbean, with taxonomic notes on the species. *The Festivus* 54(3):269-273.

Dockery, D.T. 1977. *Mollusca of the Moodys Branch Formation, Mississippi*. Bulletin 120, Mississippi Geological, Economic, and Topographical Survey. Jackson, Mississippi.

Eason, R.L. 2018. *Jaspidiconus acutimarginatus* is part of the Florida cone shell fauna. *The Festivus* 50(1):58-61.

Garcia, E.F. 2023. Species of the Families Costellariidae and Mitridae Dredged by the R/V *Pelican* in the Gulf of Mexico. *The Festivus* 55(1):65-70.

Garcia, E.F. 2023a. A new species of *Fenimorea* (Gastropoda: Drilliidae) from the eastern Gulf of Mexico. *The Festivus* 55(4):246-248.

Gustafson, R.G., R.D. Turner, R.A. Lutz, and R.G. Vrigenhoek. 1998. A new genus and five new species of mussels (Bivalvia, Mytilidae) from deep-sea sulfide/hydrocarbon seeps in the Gulf of Mexico. *Malacologia* 40:63-112.

Hildebrand, A.R., G.T. Penfield, D.A. Kring, M. Pilkington, C.Z. Antonio, and W.V. Boynton. 1991. Chicxulub Crater: a possible Cretaceous/Tertiary boundary impact crater on the Yucatan Peninsula, Mexico. *Geology* 19(9):867-871.

Marelli, D.C., W.G. Lyons, W.S Arnold, and M.S. Krause. 1997. Subspecific status of *Argopecten irradians concentricus* (Say, 1822) and of the Bay Scallops of Florida. *Nautilus* 110:42-44.

Oleinik, A.E., J.H. Leal, A. Dupont, and N. Uthairat. 2023. A prediction held true: first record of the non-indigenous Thrush Cowrie *Naria turdus* (Lamarck, 1810) (Gastropoda: Cypraeidae) in South Florida. *The Nautilus* 137(1):31-34.

Petuch, E.J. 1974. A New *Terebra* from the Coral Reef Areas off North Carolina. *The Veliger* 17(2):205-208.

Petuch, E.J. 1987. *New Caribbean Molluscan Faunas*. Coastal Education and Research Foundation, Charlottesville, Virginia. 158 pp.

Petuch, E.J. 1994. *Atlas of Florida Fossil Shells (Pliocene and Pleistocene Marine Gastropods).* Graves Museum of Natural History, Dania, Florida and Chicago Spectrum Press. 394 pp.

Petuch, E.J. 2004. *Cenozoic Seas: The View from Eastern North America.* CRC Press, London, New York, Boca Raton. 308 pp.

Petuch, E.J. 2013. *Biogeography and Biodiversity of Western Atlantic Mollusks*. CRC Press, London, New York, Boca Raton. 234 pp.

Petuch, E.J. and D.P. Berschauer. 2020. A Review of the Living *Cinctura* Banded Tulip Shells (Gastropoda: Fasciolariidae), with the Descriptions of Four New Subspecies and a New Subgenus. *The Festivus* 52(4):316-334.

Petuch, E.J. and D.P. Berschauer. 2021a. *Tropical Marine Mollusks: An Illustrated Biogeographical Guide*. CRC Press, London, New York, Boca Raton. 357 pp.

Petuch, E.J. and D.P. Berschauer. 2021b. *Ancient Seas of Southern Florida: The Geology and Paleontology of the Everglades Region*. CRC Press, London, New York, Boca Raton. 266 pp.

Petuch, E.J. and D.P. Berschauer. 2022. *Mollusks and Marine Environments of the Ten Thousand Islands*. CRC Press, London, New York, Boca Raton. 157 pp.

Petuch, E.J. and D.P. Berschauer. 2023a. A new Deep Water *Fasciolaria* (Fasciolariidae: Fasciolariinae) Tulip Shell from off the Dry Tortugas, Western Florida Keys, Florida. *The Festivus* 55(2):92-99.

Petuch, E.J and D.P. Berschauer. 2023b. New Gastropods from the Gulf of Mexico. *The Festivus* 55(4):249- 270.

Petuch, E.J. and D.P. Berschauer. 2024. New Gastropods from the Gulf of Mexico (Northwestern Florida, the Yucatan Peninsula, and the Flower Garden Reefs of Texas). *The Festivus* 56(1):18-26.

Petuch, E.J., M. Drolshagen, and G. Herndl. 2015. *Cone Shells of the Okeechobean Sea (Pliocene-Pleistocene)*. ConchBooks Publishers, Harxheim, Germany. 179 pp.

Petuch, E.J and R. Myers. 2014. *Molluscan Communities of the Florida Keys and Adjacent Areas: Their Ecology and Biodiversity*. CRC Press, London, New York, Boca Raton. 300 pp.

Petuch, E.J., R. Myers, and D.P. Berschauer. 2015. *The Living and Fossil Busycon Whelks: Iconic Mollusks of Eastern North America*. Special Publication of the San Diego Shell Club, Inc., San Diego, California. 195 pp.

Petuch, E.J. and C.E. Roberts. 2007. *The Geology of the Everglades and Adjacent Areas.* CRC Press, London, New York, Boca Raton. 212 pp.

Petuch, E.J. and D.P. Sargent. 2011. *Rare and Unusual Shells of the Florida Keys and Adjacent Areas*. MdM Publishers, Wellington, Florida. 158 pp.

Tunnell, J.W., J. Andrews, N.C. Barrera, and F. Moretzsohn. 2010. *Encyclopedia of Texas Seashells*. Texas A. and M. University Press, College Station, Texas. 512 pp.

Valentine, J. 1973. *Evolutionary Paleoecology of the Marine Biosphere*. Prentice-Hall Publishers, Englewood, New Jersey. 511 pp.

Weaver, C.S. and J.E. Dupont. 1970. *Living Volutes: A Monograph of the Living Volutidae of the World.* Monograph Series No. 1, Delaware Museum of Natural History, Greenville, Delaware. 375 pp.

Index of Illustrated Specimens

The species and subspecies listed here are arranged by genus and subgenus and are illustrated in the iconographies given at the end of each chapter, or they are inserted as fillers in otherwise blank spaces.

G

H

J

K

View from Wakulla Beach in Goose Creek Bay, "Big Bend" area along the eastern Florida Panhandle. The cordgrass (*Spartina alterniflora*) beds grows in small clumps along the low tide line and supports a population of *Littoraria irrorata sayi.*

Geographic and Oceanographic Index

The following alphabetically-arranged index lists the most important geographic, geomorphologic, oceanographic, and marine biologic terms used throughout this book.

Murexiella hidalgoi (Crosse, 1869), length 39 mm, a widespread deep water species found throughout the Gulf of Mexico and Caribbean Sea. This specimen, from 200 m depth off the Dry Tortugas, has exceptionally well-developed spines.

Close-up view of an oyster clump, at low tide on Turtle Key, Ten Thousand Islands, showing the resident ecosystem composed of barnacles (*Balanus*), bright yellow *Haliclona* sponges, black batillariid snails (*Lampanella minima*), and a rose-colored *Calliostoma tampaense*. This invertebrate assemblage is typical of those that are encountered in the back-reef lagoons found on the vermetid gastropod "reefs" (vermetoherms) of the Ten Thousand Islands of southwestern Florida.

About the Authors

Edward J. Petuch was born in Bethesda, Maryland, in 1949. Raised in a Navy family, he spent many of his childhood years collecting living and fossil shells in such varied localities as Chesapeake Bay, California, Puerto Rico, and Wisconsin. His early interests in malacology and oceanography eventually led to BA and MS degrees in Zoology from the University of Wisconsin-Milwaukee. During his Master's thesis research, Petuch concentrated on the molluscan biogeography of West Africa, traveling extensively in the Canary Islands, Western Sahara, Senegal, Gambia, Sierra Leone, Ivory Coast, and Cameroon. During this time, he also conducted research on the molluscan ecology of both coasts of Mexico and the Great Barrier Reef of Belize.

Continuing his education, Petuch studied marine biology and malacology under Gilbert Voss and Donald Moore at the Rosenstiel School of Marine and Atmospheric Sciences at the University of Miami, where he received a full scholarship. During this time, his doctoral dissertation research involved intensive field work in Costa Rica, Colombia, Venezuela, Barbados, the Grenadines, and Brazil, where he often went to sea with the local shrimpers for weeks at a time. After receiving his PhD in oceanography in 1980, Petuch was invited to conduct two years of postdoctoral research, funded by the National Science Foundation, with Geerat Vermeij at the University of Maryland. While there, he also held a research associateship with the Department of Paleobiology at the National Museum of Natural History, Smithsonian Institution, under the sponsorship of Thomas Waller, and conducted field work in the Pliocene and Pleistocene fossil beds of Florida and North Carolina and the Miocene fossil beds of Maryland and Virginia.

Petuch has also collected and studied living mollusks in Australia, Papua-New Guinea, Fijis, French Polynesia, Japan, the Bahamas, Nicaragua, and Uruguay. This research has led to the publication of over 400 scientific papers and the discovery and description of over 2,000 new species of mollusks and 200 new genera. His previous 26 books are well-known reference texts in the malacological and paleontological communities, and some of the more recent better-known ones include: *The Ecphoras: Iconic Fossils of Eastern North America* (2023); *Mollusks and Marine Environments of the Ten Thousand Islands* (2022); *Ancient Seas of Southern Florida: The Geology and Paleontology of the Everglades Region* (2021); *Tropical Marine Mollusks: An Illustrated Biogeographical Guide* (2021); *Jewels of the Everglades: The Fossil Cowries of Southern Florida* (2018); *The Living and Fossil Busycon Whelks: Iconic Mollusks of Eastern North America* (2015); *Cone Shells of the Okeechobean Sea* (2015); *Molluscan Communities of the Florida Keys and Adjacent Areas: Their Ecology and Biodiversity* (2014); *Biogeography and*

Biodiversity of Western Atlantic Mollusks (2013); *Molluscan Paleontology of the Chesapeake Miocene* (2010); *The Geology of the Everglades and Adjacent Areas* (2007); and *Cenozoic Seas: The View from Eastern North America* (2004).

Currently, Petuch is a Professor Emeritus in the Department of Geosciences, Florida Atlantic University in Boca Raton, Florida where, for thirty years, he taught undergraduate classes in oceanography, paleontology, and physical geology, and graduate classes in paleoecology and paleoceanography. He currently resides in Jupiter, Florida, with his wife Linda, where they both enjoy visits from their three children and their families.

© Morgan Taylor Photography, with permission

David P. Berschauer was born in Rockville Center, New York, in 1964, and spent his youth collecting shells in such varied localities as New York, Florida, California, Washington, Mexico, and throughout the Caribbean. His early interests in natural history, malacology, and marine biology eventually led to a BS in biology at the University of California-Irvine, an advanced marine invertebrate zoology course at Washington State University's Friday Harbor Marine Lab, and studies towards the pursuit of a graduate degree in marine biology at Florida State University in Tallahassee, Florida. While still a college undergraduate, Berschauer performed field biology research, published a number of research papers, and gave scientific presentations at national conferences.

He subsequently switched career paths and attended Southwestern University School of Law in Los Angeles, California, earning his Juris Doctorate in 1991. Although having developed a legal career spanning over thirty years and being licensed in both California and Georgia, he has kept malacology as a lifetime avocation and has put together a sizeable research collection and personal museum of molluscan specimens.

Over his entire professional life, Berschauer continued to pursue his passion for marine biology, and collecting and studying marine organisms. In his spare time, he has developed and published a relational database software program to aid in the organization and maintenance of a systematic collection. Although originally designed for malacology, the program is applicable to entomology and other aspects of systematic zoology.

Berschauer is an active member of the San Diego Shell Club, a member of the Gulf Coast Shell Club, a Museum Associate with the Natural History Museum of Los Angeles

County ("LACM") Malacology Department, a Collection Associate with the Santa Barbara Museum of Natural History ("SBMNH"), and is the Editor-in-Chief of the journal, *The Festivus*. He is also well-known for his natural history and shell photography, with a multitude of high-quality examples seen throughout this book. Besides being the author of many important scientific papers on molluscan systematics, Berschauer has described and named over 150 new species of gastropods and is also the co-author of several recent books on mollusks, including: *The Living and Fossil Busycon Whelks: Iconic Mollusks of Eastern North America* (2015); *Jewels of the Everglades: The Fossil Cowries of Southern Florida* (2018); the well-received *Sea Shells of Southern California: Marine Shells of the Californian Province* (2018); *Tropical Marine Mollusks: An Illustrated Biogeographical Guide* (2021); *Ancient Seas of Southern Florida: The Geology and Paleontology of the Everglades Region* (2021); *Mollusks and Marine Environments of the Ten Thousand Islands* (2022); and *The Ecphoras: Iconic Fossils of Eastern North America* (2023). He currently resides in Cairo, Georgia, with his wife Felicia, where they both enjoy visits with their children and their spouses, and their four young grandchildren.